Cambridge Primary

Science

Second Edition

Learner's Book 2

Series editors:
Judith Amery
Rosemary Feasey

Series authors:

Rosemary Feasey
Deborah Herridge
Helen Lewis
Tara Lievesley
Andrea Mapplebeck
Hellen Ward

AN HACHETTE UK COMPANY

Cambridge International copyright material in this publication is reproduced under licence and remains the intellectual property of Cambridge Assessment International Education.

Registered Cambridge International Schools benefit from high-quality programmes, assessments and a wide range of support so that teachers can effectively deliver Cambridge Primary. Visit www.cambridgeinternational.org/primary to find out more.

Third-party websites and resources referred to in this publication have not been endorsed by Cambridge Assessment International Education.

The audio files are free to download at www.hoddereducation.com/cambridgeextras

Acknowledgements
The Publishers would like to thank the following for permission to reproduce copyright material.

Text acknowledgements
p. 36 Song https://www.cdc.gov/features/handwashing/Public domain

Photo acknowledgements
p. 4 *cl*, **p. 59** *cc* © Koosen/Adobe Stock; **p. 4** *cc*, **p. 59** *cr* © Ilya Akinshin/Adobe Stock; **p. 4** *cc*, **p. 59** *bc* © Andrius Gruzdaitis/Adobe Stock; **p. 4** *cc*, **p. 59** *br* © Jeramey Lende/123rf; **p. 4** *cc*, **p. 26** *cl* © Yadom/Shutterstock.com; **p. 4** *cr*, **p. 9** *cl* © 2xSamara.com/Adobe Stock; **p. 4** *bc*, **p. 55** *br* © Maucorole/Fotolia.com; **p. 5** *tc*, **p. 13** *tr* © Rocket Cips/Adobe Stock; **p. 5** *cc*, **p. 13** *cr* © Andrii Zastrozhnov/Adobe Stock; **p. 8** © Fotofermer/Adobe Stock; **p. 9** *tr* © New Africa/Adobe Stock; **p. 11** *tr* © Paul Rommer/Adobe Stock; **p. 12** *bl* © Luna/adobe Stock; **p. 12** *cc* © Bajita111122/Adobe Stock; **p. 14** *cc* © Imagestate Media (John Foxx)/Vol 03 Nature & Animals; **p. 14** *tr* © Gods and Kings/Adobe Stock; **p. 14** *bl* © Vivienstock/Shutterstock.com; **p. 14** *cl* © TH Panyawachiropas/Adobe Stock; **p. 14** © Anatolii/Adobe Stock; **p. 14** *cc* © TH Panyawachiropas/Adobe Stock; **p. 14** *cc* © Dora Zett/Adobe Stock; **p. 14** *bc* © Anatolii/Adobe Stock; **p. 15** *cc* © Prostock Studio/Adobe Stock; **p. 15** *tr* © Zuzule/Adobe Stock; **p. 15** *cc* © Victoria/Adobe Stock; **p. 16** *cl* © Hogan Imaging/Shutterstock.com; **p. 16** *cc* © Rawpixel.com/Adobe Stock; **p. 16** *cc* © Ellis E. Mbeku/Shutterstock.com; **p. 16** *cr* © Monkey Business/Adobe Stock; **p. 16** *cl* © Happy Monkey/Adobe Stock; **p. 16** *cc* © Ksena32/Adobe Stock; **p. 16** *cc* © Mathom/Adobe Stock; **p. 16** *cc* © Anghi/Adobe Stock; **p. 19** *cl* © Patrick Foto/Adobe Stock; **p. 20** *tr* © Marucyan/Adobe Stock; **p. 21** *cl* © Danicek/Adobe Stock; **p. 21** *cl* © Dule964/Adobe Stock; **p. 21** *cc* © Monticelllllo/Adobe Stock; **p. 21** *cc* © Sommai/Adobe Stock; **p. 22** *cc* © Dinodia Photos/Alamy Stock Photo; **p. 22** *cc* © Lana Langlois/Adobe Stock; **p. 22** *cc* © Nipaporn/Adobe Stock; **p. 22** *cc* © Tan4ikk/Adobe Stock; **p. 22** *cc* © Tan4ikk/Adobe Stock; **p. 24** *cl* © Africa Studio/Adobe Stock; **p. 24** *cl* © Creative Family/Adobe Stock; **p. 26** *tr* © Xzotica65/Adobe Stock; **p. 27** *cl* © Konstantin Aksenov/Adobe Stock; **p. 27** *cc* © Verkoka/Adobe Stock; **p. 27** *cr* © Alexandr Mitiuc/Adobe Stock; **p. 29** *cl* © Maksym Yemelyanov/Adobe Stock; **p. 29** *bc*, **p. 144** *tr* © Sviatoslav Kovtun/Adobe Stock; **p. 29** *bl* © Dronathan Davis/Adobe Stock; **p. 31** *tr* © Lija/Adobe Stock; **p. 31** *bl* © Historic Collection/Alamy Stock Photo; **p. 31** *bc* © DPA Picture Alliance Archive/Alamy Stock Photo; **p. 31** *br* © Сергей Кучугурный/Adobe Stock; **p. 32** *tr* © Sasha Haltam/Adobe Stock; **p. 32** *bl* © Images of Africa Photobank/Alamy Stock Photo; **p. 34** *tl* © Sergey Novikov/Adobe Stock Photo; **p. 35** *tr* © TinnaPong/Adobe Stock; **p. 36** *tr* © Anish Ap1/Adobe Stock; **p. 37** *cl* © Kwanchaichaiudom/Adobe Stock; **p. 37** *cr* © WONG SZE FEI/Adobe Stock; **p. 39** *bl* © Rukanoga/Adobe Stock; **p. 40** *tl* © Protasov AN/Shutterstock.com; **p. 40** *tr* © Boule1301/Adobe Stock; **p. 40** *cl* © Delbars/Adobe Stock; **p. 40** *cr* © Anastasiia Malinich/Adobe Stock; **p. 40** *bl* © D'July/Adobe Stock; **p. 40** *br* © Archana bhartia/Shutterstock.com; **p. 41** *tr* © Asya M/Adobe Stock; **p. 41** *bl* © Quick Shooting/Adobe Stock; **p. 41** *bc* © Luiwhr/Adobe Stock; **p. 41** *bc* © Gotovyy Stock/Adobe Stock; **p. 41** *br* © Borisoff/Adobe Stock; **p. 42** *tl* © Adimas/Adobe Stock; **p. 42** *cl* © Sean Pavone Photo/Fotolia.com; **p. 42** *cr* © Graja/Adobe Stock; **p. 43** *tl* © DPA Picture Alliance Archive/Alamy Stock Photo; **p. 43** *tr* © Bhamms/Shutterstock.com; **p. 44** *cc* © Vladimir18/Adobe Stock; **p. 45** *br* © Thaweewong Vichaiururoj/Adobe Stock; **p. 46** *bl* © EB Foto/Adobe Stock Photo; **p. 48** *cl* © E Tham Photo/Alamy Stock Photo; **p. 49** *br* © Petr Malyshev/Adobe Stock; **p. 50** *cl* © Africa Studio/Adobe Stock; **p. 51** *tl* © Africa Studio/Adobe Stock; **p. 51** *tc* © Mechanik/Adobe Stock; **p. 51** *tc* © Korkeng/Adobe Stock; **p. 51** *tc* © Freer/Adobe Stock; **p. 51** *tr*, **p. 63** *tr* © Jantima plablabpo/123rf; **p. 51** *cc*, **p. 139** *tl*, **p. 141** *tl*, **p. 142** *bc*, **p. 143** © Gritsalak/Adobe Stock; **p. 53** *cc* © JC Fotografo/Adobe Stock; **p. 53** *cc* © Arcansél/Adobe Stock; **p. 53** *br* © EsOlex/Adobe Stock; **p. 53** *bc* © Arbalest/Adobe Stock; **p. 53** *bl* © Wavebreak3/Adobe Stock; **p. 54** *tr* © New Africa/Adobe Stock; **p. 54** *cl* © CRS photo/Shutterstock.com; **p. 54** *bl* © Maciej Czekajewski/Shutterstock.com; **p. 55** *cl* © Prapann/Adobe Stock; **p. 55** *cc* © Formatoriginal/Adobe Stock; **p. 55** *cl* © Showcake/Adobe Stock; **p. 55** *cc* © Photo&Graphic Stock/Adobe Stock; **p. 55** *cc* © KK/Adobe Stock; **p. 55** © Smuay/Adobe Stock; **p. 55** *cl* © Golden Leaf/Adobe Stock; **p. 58** *cl* © Lakkana/Adobe Stock; **p. 58** *cc* © Muratart/Shutterstock.com; **p. 58** *cc* © Wisunsaya/Shutterstock.com; **p. 59** *tr* © Mats Tooming/Fotolia.com; **p. 59** *cl* © Melica/Adobe Stock; **p. 60** *tr* © Africa Studio/Adobe Stock; **p. 61** © Colors0613/Adobe Stock; **p. 61** *cl* © Gavran333/Shutterstock.com; **p. 61** *cr* © Mark William Penny/Shutterstock.com; **p. 61** *cc* © Benjamart/Adobe Stock; **p. 61** *cc* © Caimacanul/Adobe Stock; **p. 64** *cr* © Hachette UK; **p. 65** *cl* © Don Bartell/Alamy Stock Photo; **p. 66** *tr* © Hachette UK; **p. 67** *tr* © The image is an illustration of Upsalite®, a mesoporous magnesium carbonate. © Disruptive Materials AB; **p. 67** *cr* © Foto Cat/Shutterstock.com; **p. 69** *tr* © Famveldman/Adobe Stock; **p. 73** *cc* © Warayoo/Adobe Stock; **p. 73** *cc* © Oleg/Adobe Stock; **p. 73** *tr* © Somchaisom/Adobe Stock; **p. 73** *cr* © Arbalest/Adobe Stock; **p. 73** *bc* © Zooropa/Adobe Stock; **p. 73** *bc* © VJ Matthew/Adobe Stock; **p. 73** *br* © Sementinov/Adobe Stock; **p. 74** *bl* © Pics Five/Adobe Stock; **p. 76** *cc* © Ecco/Adobe Stock; **p. 76** *cr* © Marcovector/Adobe Stock; **p. 76** *cc* © La Gorda/Adobe Stock; **p. 78** *cr* © Pictures News Fotolia.com; **p. 79** *tr* © Blend Images/123rf; **p. 80** *tr* © FLHC56/Alamy Stock Photo; **p. 84** *cl* © Mtaira/Adobe Stock; **p. 84** *cc* © Duncan Noakes/Fotolia.com; **p. 84** *cr* © Александр Иваченко/Adobe Stock; **p. 85** *bl* © Tony Tallec/Alamy Stock Photo; **p. 88** *cr* © Rakhmad Riyadi/Shutterstock.com; **p. 89** *tr* © Cheryl Casey/Adobe Stock; **p. 98** *tr* © Pradeep Raja Kannaiah/123rf; **p. 98** *bl* © Rakchai/123rf; **p. 100** *cc* © Thomas Dutour/Adobe Stock; **p. 100** *cc* © Nutcd 32/Adobe Stock; **p. 100** *cl*, **p. 115** *tr* © Tejmos/Shutterstock; **p. 100** *bl* © Jun SU/Fotolia.com; **p. 101** *tr* © Alswart/Adobe Stock; **p. 101** *tl* © Vvoe/Adobe Stock; **p. 101** *cl* © Yury Stroykin/Shutterstock.com; **p. 101** *cc* © Kromkrathog/Adobe Stock; **p. 101** *cl* © Titonz/123rf; **p. 101** *cc* © Tim Graham/Alamy Stock Photo; **p. 101** *bl* © Nikom Sunsopa/123rf; **p. 101** *bc* © Jrstock/Adobe Stock; **p. 102** *cl* © Steheap/Adobe Stock; **p. 102** *cc* © Mrcmos/Adobe Stock; **p. 103** *tr* © Julia Photographer/Alamy Stock Photo; **p. 103** *cl* © Agencja Fotograficzna Caro/Alamy Stock Photo; **p. 106** © Sasint/Adobe Stock; **p. 107** *cl* © Vector Tradition/Adobe Stock; **p. 109** *tr* © Daxiao Productions/Adobe Stock; **p. 111** *cl* © Logistock/Adobe Stock; **p. 112** *cc* © Morita/Adobe Stock; **p. 113** *cl* © Bluering Media/Adobe Stock; **p. 117** *br* © Monkey Business Images/Adobe Stock; **p. 121** *tr* © Creativestockexchange/Shutterstock.com; **p. 121** *cc*, **p. 131** *tl*, **p. 140** *tc* © Tyler Boyes/Adobe Stock; **p. 121** *cc*, **p. 131** *tr* © Alexlukin/Adobe Stock; **p. 121** *cc*, **p. 131** *tr* © Tyler Boyes/Adobe Stock; **p. 122** *cr*, **p. 126** *cl*, **p. 137** *tc* © Huandi/123rf; **p. 122** *cr*, **p. 126** *cl*, **p. 137** *cc* © Michal812/Adobe Stock; **p. 123** *cl* © Evgeny/Adobe Stock; **p. 125** *br* © Www3d/Adobe Stock; **p. 126** *bl* © Ekaterina/Adobe Stock; **p. 126** *cl* © Tyler Boyes/Adobe Stock; **p. 126** *cc*, **p. 137** *tr* © Fabrizio Troiani/Adobe Stock; **p. 126** *cr*, **p. 137** *cr* © Sakdinon/Adobe Stock; **p. 126** *br*, **p. 138** *tr* © Angellodeco/Adobe Stock; **p. 127** *tr* © Dmytro Surkov/Adobe Stock; **p. 127** *cl* © Zebra 0209/Shutterstock.com; **p. 127** *cr* © Jason Garnier/Shutterstock.com; **p. 128** © Whitcomberd/Adobe Stock; **p. 128** *cr* © Whitcomberd/Adobe Stock; **p. 129** *tr* © leungchopan/Fotolia.com; **p. 129** *cl* © Joel Sartore/Getty images; **p. 129** *cr* © Cathy Withers-Clarke/Adobe Stock; **p. 139** *br* © Compuinfoto/Adobe Stock.

t = top, *b* = bottom, *l* = left, *r* = right, *c* = centre
Every effort has been made to trace all copyright holders, but if any have been inadvertently overlooked, the Publishers will be pleased to make the necessary arrangements at the first opportunity.

Hachette UK's policy is to use papers that are natural, renewable and recyclable products and made from wood grown in well-managed forests and other controlled sources. The logging and manufacturing processes are expected to conform to the environmental regulations of the country of origin.

Orders: please contact Hachette UK Distribution, Hely Hutchinson Centre, Milton Road, Didcot, Oxfordshire, OX11 7HH. Telephone: +44 (0)1235 827827. Email education@hachette.co.uk. Lines are open from 9 a.m. to 5 p.m., Monday to Saturday, with a 24-hour message answering service. You can also order through our website: www.hoddereducation.com

© Rosemary Feasey, Deborah Herridge, Helen Lewis, Tara Lievesley, Andrea Mapplebeck, Hellen Ward 2021

First published in 2017
This edition published in 2021 by
Hodder Education,
An Hachette UK Company
Carmelite House
50 Victoria Embankment
London EC4Y 0DZ

www.hoddereducation.com

Impression number   10 9 8 7 6 5 4
Year   2025 2024 2023 2022 2021

All rights reserved. Apart from any use permitted under UK copyright law, no part of this publication may be reproduced or transmitted in any form or by any means, electronic or mechanical, including photocopying and recording, or held within any information storage and retrieval system, without permission in writing from the publisher or under licence from the Copyright Licensing Agency Limited. Further details of such licences (for reprographic reproduction) may be obtained from the Copyright Licensing Agency Limited, www.cla.co.uk

Cover illustration by Lisa Hunt, The Bright Agency

Illustrations by Alex van Houwelingen, Ammie Miske, James Hearne, Natalie and Tamsin Hinrichsen, Tina Nel, Val Myburgh, Vian Oelofsen

Typeset in FS Albert 17/19 by IO Publishing CC

Printed and bound by CPI Group (UK) Ltd, Croydon, CR0 4YY

A catalogue record for this title is available from the British Library.

ISBN: 9781398301610

Contents

How to use this book

This book will help you learn about Science in different ways.

Talk about what you remember or know about a topic.

What do you remember about living things?

Animals and plants are living things.
How can you tell if something is a **living thing**?

2

Look at the pictures below.

a What material is each object made of?
b Which materials are natural?
c Which have been changed?
d Which are manufactured?

Ⓐ Ⓑ Ⓒ Ⓓ

Do activities to learn more.
Work like a scientist.

Think like a scientist!

Humans are the **same** in many ways. We all have the same **body** parts. We all have a head and a body, arms and legs.

But humans are not exactly the same. We are **similar** in some ways and **different** in other ways.

Some differences are easy to see, like **height** or hair colour.

Some differences are small. You need to **observe** carefully to find them.

Learn new ideas about Science.

Talk about your ideas.

Let's talk

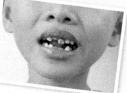

This child has tooth decay.

a Describe what his teeth look like.
b Why do you think they are like this?

Challenge yourself!

How many different cheeses can you name? Ask your family to help you.

Try something new.

Work safely! ⚠

Do not look at the Sun. It will damage your eyes.

Did you know?

Fabrics are made from different types of materials, such as cotton, wool, silk and polyester.

Learn about interesting facts and information.

Always work safely.

Science in context

What is a veterinarian?

A **veterinarian** (say: **vet**-er-in-**ar**-ian) or vet, is an animal doctor. Read the interview below to find out more about their work.

What does a vet do?

I make animals better when they are ill. I also try to stop them from getting ill. I give them tablets or injections to stop them getting fleas and worms. I help them have their babies.

What animals do you make better?

I look after people's pets, such as their cats, dogs or guinea pigs. Some vets look after farm animals, such as cows. Some vets care for unusual animals, such as zoo animals.

What helps you find out what is wrong?

I must make careful observations. I watch how the animal moves. I observe its eyes, ears, skin and fur colour. I listen to its heartbeat. I take its **temperature**.

Science words
veterinarian
temperature

Learn how we all use Science every day in our lives.

Science words
research
coir
cotton
rubber

Understand new words. The *Science dictionary* at the back of the book can also help you.

Find out how much you have learnt and what you can do.

What can you do?

You have learnt about the Earth in space. You can:
✔ use a model to show how the Sun appears to move.
✔ describe how the model appears to move in the sky.
✔ use shadows to show the Sun's movement.
✔ explain whether it is the Sun moving or the Earth.
✔ say how long a day is.

Audio links icon

Indicates content is available as audio. All audio is available to download for free from www.hoddereducation.com/cambridgeextras.

Model icon

Shows you are using a mental or physical model of something in the real world.

Star icon

Shows you need to think and work like a scientist.

Link icon

Shows you are learning things that link to another subject.

Be a scientist

What does a scientist do?

Scientists ask questions. They want to know about the world. They find answers by testing out their ideas.

Scientists use their senses. They look at what is the same and what is different in the world. They name things and sort them into groups. For example, they name the animals in the sea. They say which group they belong to.

Scientists measure things to help them answer their questions. For example, they measure how far away the stars are from Earth. They measure how cold it is from one winter to the next.

Scientists share their answers with us. They tell us what they did to find the answers.

There are different kinds of scientists. For example, geologists study rocks. Astronomers study the stars and planets.

Scientists are anywhere and everywhere. They work in schools, hospitals, businesses, in the rainforests and even on space stations going around the Earth.

Scientists are important. They help us to understand the way the world works and how to make our lives better. They discover new medicines. They create new materials. They plan how we can look after our environment.

Scientists have studied the world for hundreds of years. Some scientists are no longer alive, but they are still very famous.

You can be a scientist! Read the next page to find out how.

How do you think and work like a scientist?

Think like a scientist!

We ask questions about the world.

We talk about how to find answers.

We make **predictions** about what will happen.

We see if our predictions are correct.

We sort, **measure** and **group** things.

We **observe similarities** and differences.

We use information to answer our **questions**.

We make **models** to show our ideas.

We use **equipment** and work safely.

We use science words.

We **record** our **observations** and **measurements**.

We share our **results** with others.

Science words

predictions	questions	models	record
observations	measurements	results	equipment
observe	similarities	measure	group

What are living things?

What do you remember about living things?

Animals and plants are living things.

How can you tell if something is a **living thing**?

Science words

living thing

move   grow

food   air

humans   alive

senses

Think like a scientist!

All living things:

- **move**
- **grow**
- need **food** and water
- need **air**
- use their senses.

Humans are living things. They are **alive**. You are a human. You are alive. Plants are also living things. They are alive.

1

a Write a list of what living things can do.

b Draw three pictures of different living things.

c Under each picture write something that tells you it is a living thing.

grows

2

How many body parts can you name and point to?

This is my head.

Let's talk

Which body parts do we use for each of our five **senses**?

Same, similar, different

Think like a scientist!

Humans are the **same** in many ways. We all have the same **body** parts. We all have a head and a body, arms and legs.

But humans are not exactly the same. We are **similar** in some ways and **different** in other ways.

Some differences are easy to see, like height or hair colour.

Some differences are small. You need to observe carefully to find them.

Let's talk

What differences can you see in the faces of these children?

1

Work with a partner. **Compare** your ears. How are they the same? How are they different?

Use a mirror or a camera to see your own ear.

Challenge yourself!

Observe your partner carefully. What other differences can you see?

Science words
same   body   similar
different   compare

What body parts do animals have?

Think like a scientist!

Humans are animals. All animals have a head and a body. Many also have legs and feet.

All animals sense things around them.

Some animals have similar body parts to humans. Some have different body parts. For example, birds are animals. They have wings. Humans do not have wings.

The number or shape of animal body parts may be different.

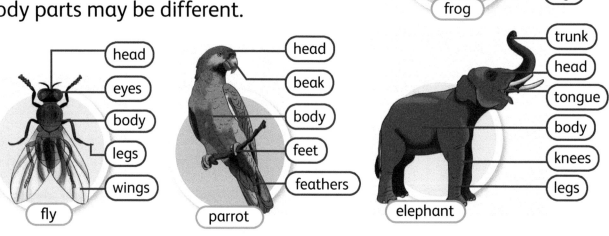

body
head
snake

eyes
skin
body
head
mouth
feet
legs
frog

head
eyes
body
legs
wings
fly

head
beak
body
feet
feathers
parrot

trunk
head
tongue
body
knees
legs
elephant

1

a Copy and complete this table. Add rows for your animals. You can use the animals on this page. One has been done for you.

Animal	Body parts:	
	Same in animals and humans	Human does not have
parrot	legs, feet, head	wings

b Talk about the differences between the body parts of animals, such as a fly and a frog.

Sort and group animals

Think like a scientist!

There are different ways to **sort** living things into groups. We can use the **features** that we can **observe** to put them in groups. Animals with similar features go in the same group.

You must be able to **describe** how you have sorted living things into groups. Then you can give the group a name.

Let's talk

It can be difficult to tell living things apart. Look at this picture. Is it a plant or an animal? What features help you to **decide**?

1

Sort these living things into groups. Give each group a name.

| parrot | spider | crocodile | whale | snake | llama | elephant |

| clown fish | frog | chameleon | hippo | tortoise | boy |

a Ask a partner to guess how you sorted the living things into groups. Is your partner correct?

b Did you both sort them in the same way?

c How many other ways can you sort them? Write a list.

Science words
sort
features
observe
describe
decide

Skin coverings

Think like a scientist!

All animals have a covering over their body. Scientists call this a **skin covering**. We can use a magnifier (say: **mag**-ni-fie-er) to look at our skin more closely. Different animals have different skin coverings. Some are covered in fur, some have feathers and some have **scales**.

Science words
skin covering
scales
moist
diagram

1

You will need...
- magnifier
- mirror

a Use the magnifier to look at your own skin. Draw what you see.
b What does the magnifier do?
c How is the skin on your arm and your face different (look in the mirror)?
d Compare your skin to your partner's skin.

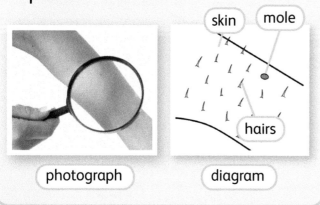

photograph   diagram

2

Find pictures of different animals. Look at their skin coverings. Sort them into these groups:

fur   scales   feathers
moist skin

Challenge yourself!

Predict which magnified animal skin covering this is.

Let's talk

Look at the **diagram** alongside. How is it different from a photograph?

Science in context

What is a veterinarian?

A **veterinarian** (say: **vet**-er-in-**ar**-ian) or vet, is an animal doctor. Read the interview below to find out more about their work.

What does a vet do?

I make animals better when they are ill. I also try to stop them from getting ill. I give them tablets or injections to stop them getting fleas and worms. I help them have their babies.

What animals do you make better?

I look after people's pets, such as their cats, dogs or guinea pigs. Some vets look after farm animals, such as cows. Some vets care for unusual animals, such as zoo animals.

What helps you find out what is wrong?

I must make careful observations. I watch how the animal moves. I observe its eyes, ears, skin and fur colour. I listen to its heartbeat. I take its **temperature**.

Science words
veterinarian
temperature

Let's talk

What science skills does a vet need? What must vets do to find out what is wrong with an animal?

Challenge yourself!

Find out who in your family uses observation in their work.

Whose baby are you?

Think like a scientist!

We know that things are living if they can eat, move, breathe, grow and use their senses. All living things also have **young** (babies). Scientists call young, **offspring**. Doctors and nurses help humans to have their offspring. Vets help animals to have their offspring.

Many offspring look like their **parents**.

Science words

young
offspring
parents

1 Match the young with its parent.

1

2

3

A

B

C

2 Some offspring have special names. Kittens are young cats. Puppies are young dogs. Kids are young goats.

a Find out the names of other young animals, such as ducks, elephants, cows, kangaroos and bears.

b Compare your names to a partner's names.

Who do you look like?

Think like a scientist!

Humans look like their parents. We do not look exactly the same as our parents. We look **similar** to our family. We are a mix of our parents. Some of our features look like our mother. Some features are more like our father.

Remember features are things such as hair colour, skin covering and number of body parts.

Many young animals look like their parents. A pony and its foal have similar markings. A goat and its kid have similar markings.

1

a Collect pictures of your family. Look carefully at them. What features can you see that are similar?

b What does it mean when someone says, "you have your mother's eyes"?

Let's talk

What parts of you are like your mother? What parts are more like your father?

Science word
similar

How do animals change?

Think like a scientist!

Offspring do not stay young forever! All young get **older** and change. All animals grow until they are **adults**.

You are **bigger** now than when you were two years old. You may be **taller**. Your arms and legs may be **longer**. Your head is bigger. You are older than you were.

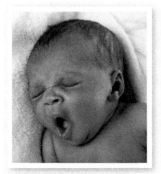   

Other animals also change as they get older.

Science words
older
adults
bigger
taller
longer

1

Look at the pictures on this page.

a How has the human changed as he grew older?

b How have the chick and the kitten changed as they grew older?

c What is similar about the young and the adult in each set of pictures?

Let's talk

How have you changed since you were a baby? What have you learnt to do? Who taught you these new things?

Who is the tallest?

Think like a scientist!

Scientists say that as we get older, we get taller. This is a **pattern** that human beings follow. We can recognise a pattern when we observe that as one thing changes, we know what change will happen next.

Science word
pattern

1

a Stand next to a partner. Who is older? Who is taller?

> I am 6 and a half years old. I'm taller than you.

> I am 6 years old.

b Is there a pattern of height and age in your class? Make a prediction.

c How could you find out if your prediction is correct?

d Try it. What did you find out?

2

a Read the *Science words* that end in -er on page 16. We use words like these to compare things, such as big and bigger, tall and taller.

b Write your own list of words you find that end in -er.

c Write the opposite of each of the words in your list.

What have you learnt about humans and animals?

1

a Collect pictures of animals.

b Challenge a partner to sort them into groups.

c Guess how they have sorted the animals.

d Swap over. Get your partner to guess how you have sorted them.

2

Name two ways you have changed as you have grown older.

What can you do?

You have learnt about humans and animals. You can:

✔ make careful observations.

✔ compare features of different people.

✔ use a magnifier.

✔ compare different animals.

✔ name some features that we use to group animals.

✔ group animals based on their features.

✔ recognise different types of skin coverings.

✔ describe why it is important to make careful observations.

✔ describe how you and other animals change as you get older.

✔ recognise that animals have offspring that look like their parents.

What do living things need?

What do you remember about being healthy?

Scientists say we need to eat lots of different foods to be **healthy**. Scientists call everything we eat, our **diet**. Eating is one of our life processes. What other **life processes** can you remember? Make a list.

Think like a scientist!

All living things need water. When you feel **thirsty**, your body is saying, "Drink water!"

Scientists say we should drink eight glasses of water every day. This helps us to think clearly and stay healthy.

Let's talk

How many different words can you use to say you are **hungry**? Sort your words. Use these groups to help you:

| most hungry | | hungry |
| very hungry | | least hungry |

1

Answer these questions.

a What makes you feel thirsty?

b What do you drink to get rid of your thirst?

c What drinks do you like?

Did you know?

More than half of your body is made up of water!

Science words

healthy   diet   life processes
thirsty   hungry

Fruits, vegetables and water

Think like a scientist!

Our diet must have foods from different **food groups**.
Each food group gives us something different to help us stay healthy.

Fruits and **vegetables** are one of the food groups. Scientists say we should eat five fruits or vegetables a day to be healthy. Many fruits and vegetables also give us **water**.

1

Draw a basket.

a Inside your basket, draw different fruits and vegetables that you like to eat.

b Show your drawing to a partner. Do you both like the same fruits and vegetables?

c Do you eat five lots of this food group a day? How could you find out?

Science words

food group   fruit
vegetable   water

2

Look at the picture of the basket of fruits and vegetables above.

a Which fruit or vegetable have you never tried?

b At home, ask if you can try some to find out if you like it.

Challenge yourself!

How much water do you drink every day? Make a prediction.

Record what you drink. Don't forget to include foods that give you water too! Was your prediction correct?

Bread, rice, pasta and potatoes

Think like a scientist!

Scientists say we should have three meals a day. Breakfast is the most important meal in our diet. It keeps us moving all day.

Bread, rice, pasta and potatoes are part of the food group that helps us to move around and play. This food group is good for breakfast.

rice   pasta   bread   potato

1 ⭐

a Which of the foods above do you eat most of? Make a prediction.

b Keep a food diary for one week and write down all the rice, pasta, bread and potatoes you eat.

c Draw a block graph to show what you found out.

d Use your block graph to work out which of the foods in the box above you eat most of. Was your prediction correct?

2

What do you notice when you eat these foods? Do you eat them on their own or with something else? What else do you eat them with?

Do you think it would be good to eat only pasta? Why?

Let's talk

Talk about the meals you ate yesterday. Which foods do you eat most of in each meal?

Dairy foods

Think like a scientist!

Dairy foods are made from milk. Milk comes from animals such as cows, sheep and goats.

Dairy foods help our bones and teeth to grow **strong**.

yoghurt   milk   different cheeses

1

Look at the dairy foods above.

a Which of these dairy foods do you eat?

b What other dairy foods do you like to eat?

c What flavours of yoghurt do you like? Try some at home to find out.

d Draw a picture of the dairy foods you eat.

Challenge yourself!

How many different types of cheese can you name? Ask your family to help you.

2 ⭐

Some people cannot eat dairy foods. What do they eat to help their bones and teeth to grow strong? They may drink a liquid from nuts, such as almonds, called almond milk. Research what other foods they should eat.

We drink milk. There is another important liquid we drink as well. What is it?

Science words

dairy   strong

Meat, fish, eggs and beans

Think like a scientist!

Meat, **fish**, **eggs** and **beans** help to keep our muscles, eyes and skin healthy. We need our muscles to help us move. Our skin protects our body. Our eyes allow us to see. This group of foods also helps to stop us from getting ill.

1

Look at these different foods.

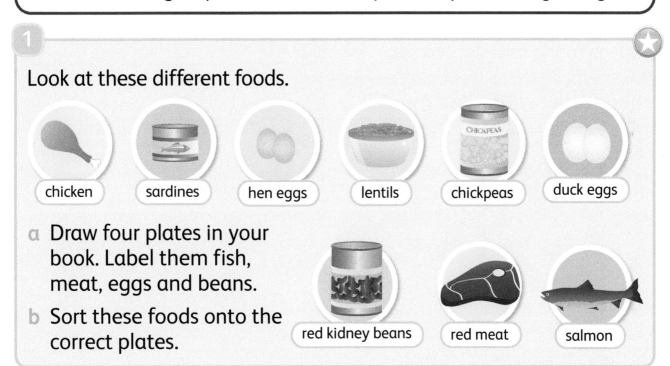

chicken   sardines   hen eggs   lentils   chickpeas   duck eggs

red kidney beans   red meat   salmon

a Draw four plates in your book. Label them fish, meat, eggs and beans.

b Sort these foods onto the correct plates.

2

Plan your own healthy meal.

a Use all the ideas scientists have given us to make a picture model of a healthy meal. Use the paper plate.

b Ask a partner to check your healthy meal:

  • Does it have food from each food group?

  • Is this a meal you would eat at home?

  • What would you drink?

You will need...
  • paper plate

Science words
meat   fish
eggs   beans

Which are your favourite foods?

Think like a scientist!

The cakes and sweets in the pictures below look delicious! But they have a lot of **sugar** in them. Too much sugar is bad for our teeth and bodies.

Some foods have lots of **fat** in them. Too much fat can be bad for our health. We should not eat too many of these foods.

cakes and biscuits have lots of fat and sugar

sweets have lots of sugar

Science words

sugar	fat
survey	tally

1

a Draw your favourite meal. Label each food in the meal.

b Is your meal healthy? How do you know?

2

a Predict what food learners in your class would like most.

b Carry out a **survey** of your class to find out.

c Copy and complete this table to record what you find out. Write in your own foods and **tally** the numbers.

Food	Tally	Number
apples	𝍤 I	6
chocolate		
chicken		

d Look at your results.

- Which was the favourite food in your class?

- Was your prediction correct?

- Which food did learners like the least?

Healthy eating traffic light

Think like a scientist!

We must choose what we eat to stay healthy. Foods that have too much fat and sugar can be bad for our health. These are *red-light foods*. We can have a small amount of these foods. It is healthier to choose another food.

Every meal should have foods that help us grow. We do not need a lot of these. They also help our body to get better or heal when we are ill or injure ourselves. These are *yellow-light foods*.

Every meal should have lots of foods that help us move and play. These are *green-light foods*. They also make sure that our body works well.

1

Look at the two lunchboxes.

a **Classify** (sort) the foods in each lunchbox into red, yellow and green foods.

b Are the lunchboxes a **balanced** meal?

c Mix the foods in the lunchboxes. Make new, **healthier** lunchboxes.

egg   sweet
cheese   chicken
starfruit banana
grapes   carrots

lollipops   crisps
sweets   sandwich
cupcakes   soda

Science words

classify
balanced
healthier

Tooth decay

Think like a scientist!

Our teeth can rot and get holes if we eat food and sweets that have a lot of sugar. Scientists call this **tooth decay**. **Dentists** help us look after our teeth. They call a hole in our teeth, a **cavity**. If you see a black spot on a tooth, this could be a cavity starting. A dentist may give the tooth a **filling**.

Sugar can also cause **gum disease**. This is when your gums feel sore and may bleed.

Let's talk

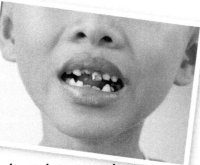

This child has tooth decay.

a Describe what his teeth look like.

b Why do you think they are like this?

Science words

tooth decay   dentist
cavity   filling   gum disease

1

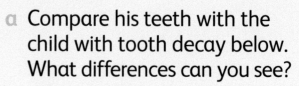

This teenager does not have tooth decay.

a Compare his teeth with the child with tooth decay below. What differences can you see?

b Which teeth would you rather have?

2

a Which of these things do you think the teenager ate or used to keep his teeth healthy?

b What does your family do to keep their teeth healthy?

biscuit   fizzy drink

chocolate   pasta

toothpaste   cake

cheese   milk

toothbrush

How many teeth?

Think like a scientist!

When we are born, we do not have any teeth. At about six months old we get our first teeth, called **milk teeth**. You should have about 20 milk teeth now. You will not keep these teeth. Adult teeth grow underneath the milk teeth. They push the milk teeth out. Adults have 32 teeth.

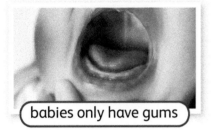
babies only have gums

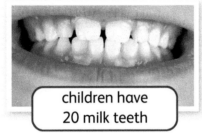
children have 20 milk teeth

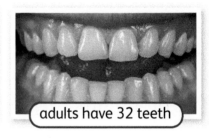
adults have 32 teeth

1

You will need...
- mirror or a digital camera
- worksheet of a model mouth

a Look in the mirror and smile.

b Draw a picture or take a photograph of you smiling and showing your teeth.

c Open your mouth wide and look at your teeth.

d How many teeth do you have?

e Do you have any fillings or tooth decay?

f Complete the worksheet your teacher gives you. Label your teeth and colour in your fillings. Show where a tooth is missing.

Let's talk

When did you lose your first tooth? How many teeth have you lost? Do you have any adult teeth?

Science word
milk teeth

Why are teeth important?

Think like a scientist!

We have four different types of teeth. Each tooth is important and has a different job. They have different shapes to match their job.

We use sharp **incisors** (say: in-**size**-ers) to cut pieces off our food.

Our **canines** (say: **cay**-9s) help to grip the food as we bite it. They are pointed to dig into our food.

Our **premolars** (say: **pree**-mole-arrs) and **molars** (say: **mole**-arrs) are flat on top. They **grind** and **chew** our food. Chewing makes the food small enough to **swallow**.

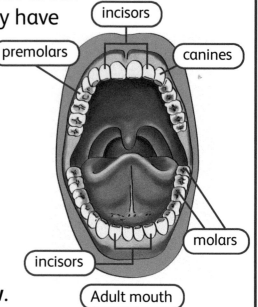

incisors

premolars

canines

molars

incisors

Adult mouth

1

a Read the *Think like a scientist!* again.

b Look carefully at the teeth in your mouth. How many of each tooth type do you have?

2

a Add the names of the teeth types to your worksheet of a model mouth with 20 milk teeth.

b Is this model a diagram or a picture? How can you tell?

Challenge yourself!

Take a bite of food such as an apple. Chew it. What is the name of the teeth that you used to bite the food? Can you use other teeth to bite the food? Try it.

Science words

incisor	canine
premolar	molar
grind	chew
swallow	

Look after your smile

Think like a scientist!

There are lots of ways to help look after your teeth and your smile. You should not eat foods with lots of sugar in them. **Brush** your teeth twice a day. Brush for two minutes with a **toothbrush** and **toothpaste**.

The dentist will check if your teeth have any decay. The dentist will ask if you are brushing your teeth properly.

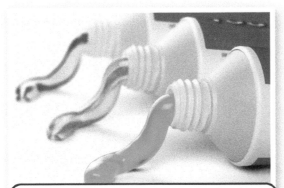

Toothpaste helps to make teeth white, protects your gums and stops tooth decay.

Some toothbrushes use electricity to make them move. They have a cell (battery).

1

a Which type of toothpaste does your class use most? How can you find out?

A group of learners recorded their favourite toothpaste. Here are their results:

Type of toothpaste	Number of learners who like it
type 1	6
type 2	15
type 3	9
type 4	10

b Draw a **block graph** of their results.

c Which toothpaste did the most learners like?

d Make a table and a block graph with the results from your class.

Science words
brush   toothbrush
toothpaste   block graph

Science in context

How to clean your teeth

We must brush our teeth to keep them healthy and clean. Dentists have some simple rules to follow:

- Brush your teeth with toothpaste at least twice a day.

- Also brush after you eat anything. This is not always possible, but try!

Use a brush the right size for your mouth.

Only use a small amount of toothpaste. Wet it with water.

Brush gently in small circles over all your teeth on the inside and outside, for two minutes.

Spit out the toothpaste. Do not rinse your mouth out. This will remove all the toothpaste!

2 minutes

1

Look at the leaflet for brushing teeth above.

a How long should you brush your teeth for?

b How many times a day should you brush your teeth?

c How much toothpaste should you use?

d Should you rinse your mouth after brushing?

2

Make a video with two other learners. The video should teach younger children how to clean their teeth.

a Plan your video. Practise acting out how to clean your teeth. Work out what you will tell the children.

b Make the video. Choose one learner to act, one to talk and one to take the video.

c Show your video to some younger children. Do they like it?

Plaque causes tooth decay

Think like a scientist!

There is a special tablet that shows if we have brushed our teeth properly. It makes the dirt or **plaque** left on our teeth turn a different colour.

Plaque is a sticky, slimy, creamy colour. It causes tooth decay.

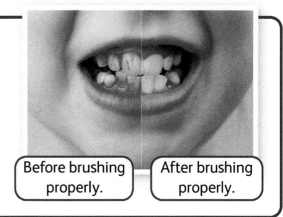

Before brushing properly.

After brushing properly.

Let's talk

Talk about the red plaque in the picture. Where is most of the plaque? Why do you think this is?

Science words
plaque
false

Did you know?

Many years ago, people did not know how to care for their teeth. Many people had tooth decay. Their teeth fell out. Sometimes they used pliers to pull their teeth out – how horrible!

Some people had new **false** teeth made.

false teeth from over 200 years ago, made of leather, metal and other people's teeth

false teeth from 200 years ago, made of hippo and elephant tusks, and gold

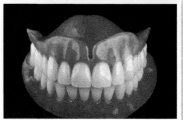

modern false teeth made of plastic

Active and inactive

Think like a scientist!

We can stay healthy by brushing our teeth and eating the right foods. We also need to **exercise** to be healthy. When we run, skip, hop, dance, play football or other sports, we are exercising. We are being **active**. This keeps our body strong and healthy. It is good for our **muscles**. How active we are is called our **lifestyle**.

1

Which of these activities are more active than others? Sort them into an activity list – from very active to **inactive** (opposite of active).

very active ⟵ active ⟶ inactive

2

Some people play tennis. Some like to paddle a canoe.

a What is your favourite exercise? How does it keep you healthy?

b Use the words in the *Science words* box in your answers.

Challenge yourself!

Find out what your family does to exercise. Is their lifestyle active or inactive?

Science words

exercise	active
muscle	lifestyle
inactive	

A day with Jack

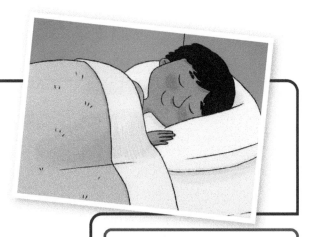

Think like a scientist!

Scientists have found that **sleep** is as important as healthy eating and exercise. A seven-year-old should sleep for about 11 hours each night.

Scientists say that too much **screen time** makes you inactive and this is not healthy.

Science words
sleep   screen time

1

Jack recorded what he did for a day. He made a block graph of his results.

a Which activity takes up most of the day?

b Which activity does Jack do for the shortest amount of time?

c Which activity could Jack do less of or swap for a healthier activity?

d Do you think Jack is healthy? Why?

e How could he make his lifestyle healthier?

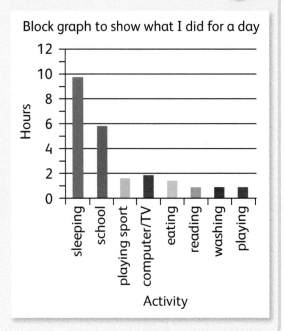

Block graph to show what I did for a day

2

a Create a list of what you do each day. Compare Jack's list to yours. What is the same? What is different?

b Think of one thing that you can do to make your life healthier.

c Create a block graph like Jack's to show the results of your day.

How to keep healthy and happy

Think like a scientist!

Exercise is good for us! If we have a strong and healthy body that we can bend, we will be able to run, jump, skip and do all the sports we like. Scientists have found that people who run and jump are happier than people who do not exercise!

1

When you eat well and exercise enough, your body works well, and you feel good. This is the same as saying that you are healthy and happy. You can do all the things you want.

Remember these things to stay healthy:

- Eat a balanced diet.
- Drink lots of water.
- Get enough sleep (at least 10 hours).
- Only watch television or a computer screen for a short time.
- Move around and exercise.

Write an idea to help you remember each point. For example: I will try to eat more vegetables. I will eat cake only on special occasions.

Do you remember the healthy eating traffic lights? Look at page 25.

Let's talk

How does exercise make you happy? Make a list of the reasons to do exercise.

Keep clean

Think like a scientist!

Our hands may look clean, but are they? We may have millions of **germs** on each hand! Some of these germs are harmless. Some can cause **illness**.

To stay well and healthy we must be **hygienic** (say: hi-**jean**-ic). It is hygienic to wash our hands with soap and water. This helps stop the spread of germs.

1

Match the problem to the answer.

Problem	Answer
a Damp hands spread 1000 times more germs than dry hands.	1 Use soap to wash your hands.
b The number of germs on our fingertips doubles after we use the toilet.	2 Dry your hands properly after washing them.
c Water does not kill the germs on our hands.	3 Wash your hands after every visit to the bathroom.

2

a When do you wash your hands – before or after an activity?

b Use these groups to sort out when you wash your hands:

before   after   before and after

c Draw a block graph to show if you wash your hands more before an activity or after.

Science words
germs   illness
hygienic

The hand washing rule

Think like a scientist!

We must wash our hands for at least 20 seconds. Doctors and vets wash their hands very carefully for at least 20 seconds to help stop the spread of germs. We must follow this hand washing rule.

Did you know?

20 seconds is as long as it takes to sing the *Happy Birthday* song twice! Try it the next time you wash your hands.

1

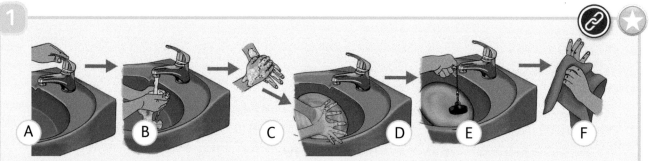

A   B   C   D   E   F

This flow diagram shows how to wash your hands. Write an instruction for each stage.

2

a There are many different soaps we can use to wash our hands. Find out the name of some soaps your class uses.

b Look at the different types of soap you have at home. How many different kinds do you use?

Did you know?

We use special soaps to clean the kitchen, bathroom and floor. They all help to kill germs.

Stop illnesses!

Think like a scientist!

Many years ago, humans did not stay as clean as they could. We had many more illnesses than we have today. Scientists now know that to be healthy and well, we need to be clean.

We can become ill if germs are passed from one person to another.

Different illnesses make us feel ill in different ways. We will have **symptoms** that tell us what illness we have.

Science words

symptoms   chickenpox

1

a Read about three illnesses.

If someone sneezes or coughs on us, their germs can give us a cold. This makes our nose run. We may get a sore throat. We may have a headache. We should put our hand or tissue in front of our mouth and nose when we cough or sneeze.

Some illnesses can cause a rash. We will have spots. **Chickenpox** spots are red. They can be itchy. We should try not to scratch them. We feel hot.

If we eat food with unclean hands, we could get a tummy ache, be sick or go to the toilet a lot. Unclean hands have 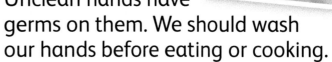 germs on them. We should wash our hands before eating or cooking.

b Draw a table like this one. Write the illness and its symptoms from the sentences above.

Illness	Symptoms

What have you learnt about being healthy?

1

Which of these foods should you only eat every now and then?

A  apple

B cheese

C  naan bread

D cake

E 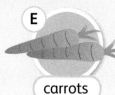 carrots

2

a Do these foods give a balanced, healthy meal? Why?

b Sort the foods into the food groups that you learnt about.

c Sort the foods using the healthy eating traffic lights on page 25.

d What food or drink could you swap for healthier options? Explain your ideas.

 fizzy drink

 chicken curry with rice

baked pineapple

naan bread

What can you do?

You have learnt about being healthy. You can:
- ✔ say what kinds of food groups we need for a healthy and balanced diet.
- ✔ say why we need water to live.
- ✔ explain what an illness is.
- ✔ describe why teeth are important and how to care for our teeth.
- ✔ explain why hand washing is important.
- ✔ say why we need to exercise to stay healthy.
- ✔ record and present information you have found.

Living things

What do you remember about living things?

Humans, other animals and plants are living things. All living things need somewhere to live. They need a home.

Draw a picture for one of the life processes of a living thing. Share your pictures with other learners.

move

1

a Read this scientist's report on a new discovery called a tardigrade (say: tar-**de**-grade).

b Find all the things in the report that tell you if the tardigrade is a living thing or not.

c Which words tell you where the tardigrade lives?

Science words

discovered

shallow

This strange looking animal is called a tardigrade. It is only 1 mm long. It was **discovered** in Japan in a pile of moss. The tardigrade eats the juices from the moss it lives on.

The tardigrade can move in two ways. It has eight legs to walk. It can also swim, as it is found in **shallow** water. The tardigrade doesn't have any eyes. It can sense light and the movement of other tardigrades. If there isn't enough air around the tardigrade, it will flatten out so that it can breathe.

Living things live in different places

Think like a scientist!

Plants and animals can be found in different places all over the world. Plants live where there is light and water. Animals live where there is food and water.

There are living things in dry places and in wet places.

scorpion

frog

Animals and plants live in hot places and in cold places.

elephants

reindeer

Animals and plants live on mountains and in rivers.

mountain goat

fish

Let's talk

Look at the pictures above. Describe where each living thing is found. Talk about other places where animals and plants live.

What is the environment?

Think like a scientist!

Where animals and plants live is called the **environment** (say: en-**vie**-ron-ment). An environment is made of the weather and **landscape** that the animal or plant lives in. A landscape is all the things around us, such as trees, hills and buildings. A mountain is an environment. So is a forest.

Not all environments are the same. Some plants and animals can only live in certain environments.

The penguins in the above picture live in the Antarctic. There are very few plants here. It is very cold and snowy. The penguins swim in the sea to find food. They eat fish. Penguins need to live where they can catch fish.

1

Look at the different environments below.

a What do you think lives there?

b Make a list. Remember that plants are living things!

c Find out what lives there. Check your ideas. Were any of your ideas correct?

d What was the most unusual living thing you found there?

Science words
environment
landscape

tropical forest

polar regions

desert

coral reef

Places to live

Think like a scientist!

Science word
habitats

Living things need a place to live. Different living things live in different environments. There are special places in the environment where they live. They are called **habitats**. Habitats can be very big or very small.

▲ A daisy lives in a grassy meadow. A meadow is a large habitat.

▲ The shark lives in the sea. The sea is a very big habitat.

▲ The hairy caterpillar lives on leaves. A leaf is a small habitat.

1

First think about the environment you live in.

a   Describe the land. Is it flat or hilly? Is it grassy or sandy?

b   Describe the weather. Is it hot or cool? Does it rain a lot or is it very dry?

Now think about your habitat. This is your home.

c   Draw a picture of your home and show it to your partner.

d   Describe how your home gives you everything you need.

Science words
explorer   graphs
experiment

How do we find new living things?

Scientists and **explorers** find new living things all the time.

▲ This tiny chameleon was discovered in 2012 in a forest in Madagascar.

▲ This frog was discovered in a marsh in New York City in 2008.

Scientists ask questions and try to find answers.

Scientists make observations. They carry out **experiments**. They record their work. They take measurements and draw diagrams and **graphs**. Scientists write up their conclusions and share them with us.

1 Do you think an explorer is a scientist?

Explorers go on journeys. Some journeys are long. Some are short. Explorers want to find answers to questions.

Sometimes explorers find new living things. Sometimes they find new places. Sometimes they learn more about things they already know.

Explorers observe what they find. They record their journeys. They may keep a diary. They draw diagrams, pictures and maps. They take photographs.

Explorers share what they have found out. This helps us to understand the world around us.

Record living things

Science word

image

Think like a scientist!

Scientists and explorers observe and record what they discover. This helps us to learn about and share what they have found. Sometimes we can show others the real thing. Sometimes we can share a picture or image. An **image** is a model of the real thing. Some images are better at giving information than others.

1

A group of learners were explorers. They went around their school to find living things in their habitats. They found a butterfly. They recorded it in three different ways.

A 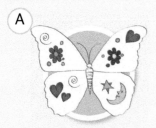 B  C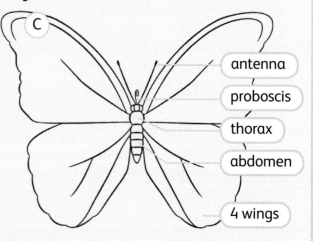

antenna

proboscis

thorax

abdomen

4 wings

Look carefully at what these learners recorded.

a What is similar about all three images?

b One image is a picture, one is a diagram and one is a photograph. Which is which?

c Describe the difference between the picture and the diagram.

d Which image do you think gives us the most information?

A photograph is an image and is not real. It is a model.

A diagram is a black and white drawing with labels.

Be an explorer!

Think like a scientist!

There are different habitats in an environment. Explorers go on safari (a hunt) to find living things in their habitats.

1

Be an explorer and go on safari.

a Walk around your school with a partner.

b Decide on a habitat to explore. This could be a tree, under a bush or near some water. Lift **stones**. Look under bushes. Look in the leaves.

c Record what the habitat is like. Use any of the ways in the *Let's talk* activity.

d Record all the living things that you can see in the habitat.

Let's talk

How can you find out what is living near you? How would you record your observations, if you were an explorer? Would you:

- use a camera?
- draw diagrams with labels?
- draw pictures?
- use words?

What else could you do?

Work safely! !

Place things back where you found them.

Wash your hands after you touch any animals, plants or **soil**.

Science words
stone   soil

Habitats in an environment

Think like a scientist!

A habitat is a special place where a plant or animal lives. A habitat has these things:

- **shelter**
- food
- air
- place to care for offspring
- water.

A pond is a habitat. The duck and its ducklings can shelter next to the pond.

The pond has small fish, snails and plants for the ducks to eat.

The pond has water for them to drink.

There is air for them to breathe.

1

Science word
shelter

a Make a model habitat for ducks. Use any material you can think of. Here are some ideas:

tray   grass   water   plastic

ducks   dough   soil   pebbles

b Could a duck live in your model? Why?

2

Match these animals to their habitats.

bear

soil

river

worm

cave

insect

fish

woodpile

Who lives where?

Think like a scientist!

Animals and plants live in places that match their needs. There are areas in the local environment that some animals and plants prefer. These are their habitats. **Invertebrates** (say: in-**ver**-te-brits) are very small animals that live in small habitats. They prefer to live in some places more than others.

1

Noor and Sara went on an invertebrate safari. They made this block graph.

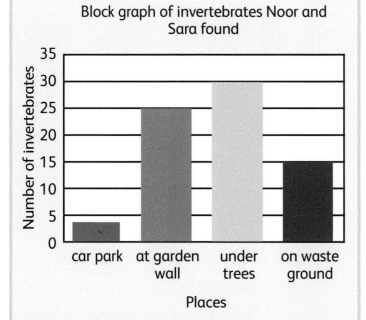

Block graph of invertebrates Noor and Sara found

Number of invertebrates (y-axis: 0, 5, 10, 15, 20, 25, 30, 35)

Places (x-axis: car park, at garden wall, under trees, on waste ground)

a Where did Noor and Sara find the most animals?

b Why do you think there were fewer animals in the car park?

2

a Make a similar block graph of the living things you found on your safari in Activity 1 on page 45.

b Where did you find the most animals?

c Compare the animals you found with those of other learners. Are they the same or different?

d Why are some different? Were they in different habitats?

Did you know?

There are about ten quintillion insects in the world. That is 10 with 18 zeros! Try writing this number down.

Habitats match needs

Think like a scientist!

Would a polar bear survive in the desert?

Have you ever seen a whale living in a tree?

All living things live in their own special habitat that matches their needs.

Some habitats are tiny, such as under a stone. These tiny habitats are called **micro-habitats**. All habitats have things that animals or plants need to live.

micro-habitat of ants

Let's talk

Talk about your habitat or home.

a What does your habitat or home have that you need to live?

b Share your ideas with another pair. Are your ideas the same or different?

1

Did you find any micro-habitats on your safari in Activity 1 on page 45?

Science word
micro-habitats

Everything you need

Think like a scientist!

Your home is a habitat. You are an animal. You have needs like other animals. Your habitat is a place that gives you everything you need to live.

1

a Maris started a chart to show the things her home has. Start your own chart to show the things your home has.

b What would happen if you had none of these things? Would you survive?

c Think of an animal you found on your safari. Fill in a chart for that animal. Does it need the same things as you?

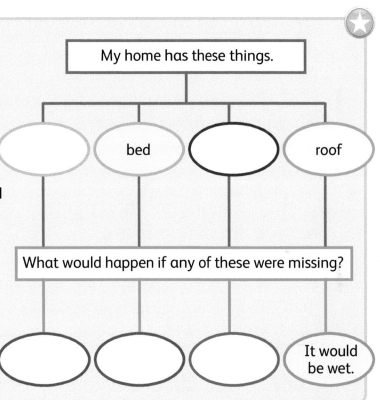

My home has these things.

bed

roof

What would happen if any of these were missing?

It would be wet.

2

a Draw and label a picture of your perfect home.

b What special things would you have in it? Which things are just nice to have? Which things do you need to live?

Some animals carry their homes around with them, like the hermit crab.

hermit crab

What have you learnt about environments?

1

a Work as a group. Write a quiz that has five questions. The questions should test what you know about living things in the environment. Make sure you know the answers.

b Give your quiz to another group to answer.

c How many questions did they get correct?

2

a Pick an environment. Describe the clothes you would need to wear there.

b Describe the animals and plants that you could find there.

c Can others guess where you are?

What can you do?

You have learnt about environments. You can:

✔ name some living things around you.

✔ recognise that an image is a model of something.

✔ say how you understand more about the environment now.

✔ decide how to collect and record observations.

✔ describe what a habitat is.

✔ describe some habitats and say what lives there.

✔ use different sources to find out about the environment.

1 Which of these are living things?

rabbit   car   snake   cactus   flower   chair

2 Draw a healthy meal on a plate. Draw a healthy drink next to it.

3 Match the tooth to the correct description below.

Ⓐ   Ⓑ   Ⓒ   Ⓓ

canine   incisor   premolar   molar

a These teeth are used to grind food to make it smaller.

b We can grip and tear food with these teeth.

c These are the biggest teeth. They grind up food.

d These teeth bite off pieces from the food.

4 Match these foods to why we need to eat them.

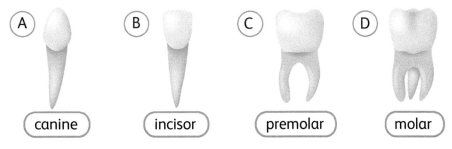

Ⓐ meat fish eggs beans

Ⓑ fruit vegetables

Ⓒ milk cheese yoghurt

Ⓓ bread rice pasta
    potatoes

a Keeps us healthy.

b Gives us energy.

c Helps bones and skin be healthy and repair itself. Stops us getting ill.

d Helps us grow strong bones and teeth.

5 Describe how you have changed since you were a baby.

6 Where does each animal live?

desert habitat

rainforest habitat

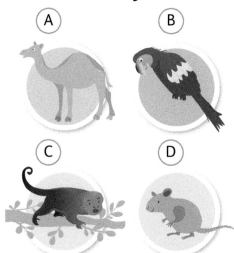

7 Some learners looked for small animals. This is their graph.

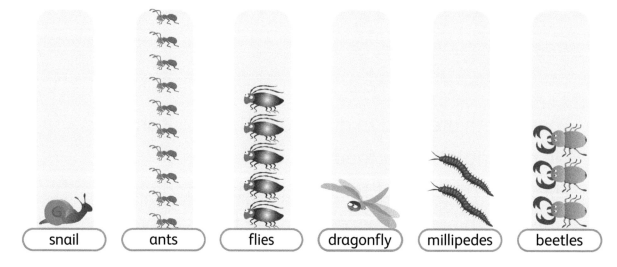

snail   ants   flies   dragonfly   millipedes   beetles

   a How many animals did they find in total?
   b Which animal did they find the most of?
   c Name another animal they might find.

8 What information does a diagram give you that a photograph or picture does not?

9 Copy and complete this sentence.

   An animal's home is called its _____ .

Materials

What do you remember about different materials?

How many **materials** can you remember? We will find out more about materials and what we can use them for.

1

a How many different materials can you see here? Write a list.

b Choose three of the materials in your list. Write one word to describe each one.

2

Some learners have not put away their toys. Help them to sort the toys below by the type of material – plastic, metal, wood or fabric.

a List the toys for each box.

b Compare your list with a partner's. Are they the same?

3

We can change the shape of some materials. List the ways we can do this.

Science word
materials

53

Materials in your classroom

Think like a scientist!

Everything is made from materials. All the things we use, our clothes and our toys are made from materials. You already know the names lots of materials. Scientists call things made of materials, **objects**. An object can be made of more than one material. A cardigan is made of wool. It also has plastic buttons.

metal

wood

cotton

plastic

wool

plastic

Science word
objects

Let's talk

Look at your chairs and desks. What are they made of?

- Are they made of wood?
- Are they made of metal and plastic?
- Are they made of other materials?

1

Look around your classroom. What other objects can you see that are made of more than one material?

Draw a table like this one. Record what you find.

Object	Materials
chair	wood and metal
window	glass and...

a Which material is used most?

b What two materials are used together the most?

c Are there any materials that are not used together?

Sort materials

Science words
fabric   classifying

Think like a scientist!

Scientists sort and group things. They call this **classifying**. We can use sorting circles for our groups.

1

Practise grouping.

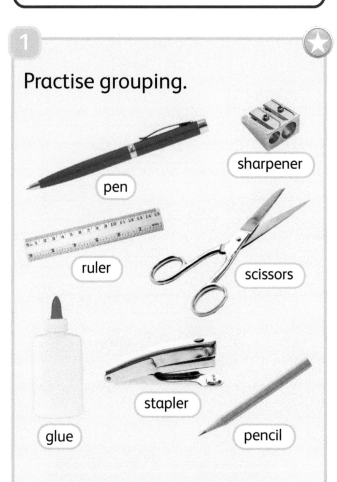

pen

sharpener

ruler

scissors

glue

stapler

pencil

a Draw three sorting circles.

b Sort the objects in the pictures into the sorting circles. Write them in.

c Label each group with the material name.

2

Test your partner!

a Look at the table you drew in Activity 1 page 54.

b Count how many different materials you found.

c Draw a sorting circle for each material type. Do not label the circles.

d Sort the objects into the sorting circles. Write them in.

e Ask your partner to guess the name of each group.

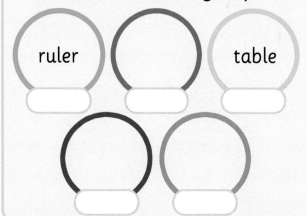

ruler

table

Did you know?

Fabrics are made from different types of materials, such as cotton, wool, silk and polyester.

Materials graph

Think like a scientist!

We can share our results in different ways. Scientists often draw tables or block graphs to share their results.

1

A group of learners looked around for objects made of different materials. This block graph shows how many objects they found.

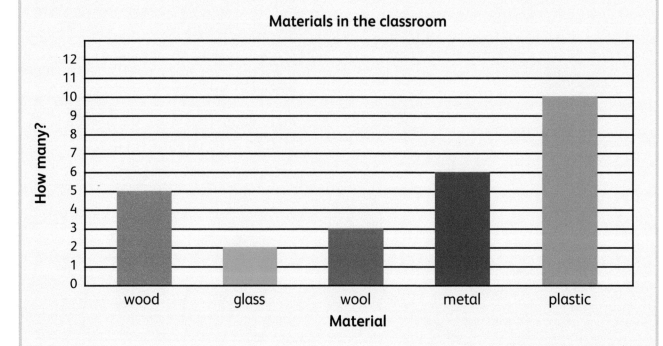

a Copy this table. Use the block graph to complete the table.

b Which material did the learners find most of in their classroom?

c Why do you think there were only two glass objects?

Material	How many
wood	
glass	
wool	3
metal	
plastic	

Where do materials come from?

Think like a scientist!

Science words
natural   manufactured

Materials come from different sources (places or things).

We find some materials in nature, such as wood and rock. They are **natural**. Some materials are made by people. We call these materials **manufactured**. People make things like plastic, glass and paper.

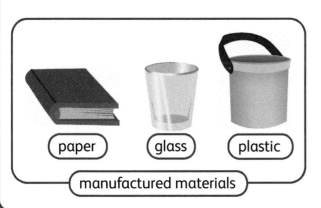

paper   glass   plastic

manufactured materials

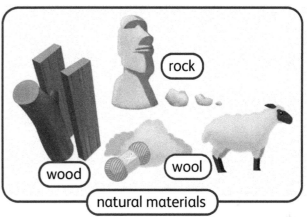

rock

wood   wool

natural materials

1

You have been a plant explorer before. Now be a materials explorer! Take a notebook and pencil with you.

a Walk around your school. Look for different materials.

b Do not damage anything. Draw the materials that you see.

c Talk to your group. Which materials look natural? Which are manufactured?

d Sort your materials into two groups – natural and manufactured.

e Record your items into two sorting circles like these.

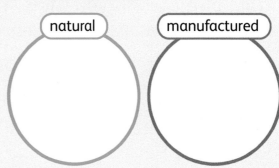

natural   manufactured

Materials from plants

Think like a scientist!

Some of the natural materials we use come from plants.

Wood comes from trees. Did you know that we use materials from plants in different ways?

1

Do **research** with a partner. Find out which of the three plants changes into which material.

A

B

C

coir mat

rubber gloves

cotton shirt

2

a Look around your school or home. Write a list of materials you think come from plants. How are these materials used?

b Compare your list with your partner's list. Check with your teacher if you are unsure about any material.

c How many did you find?

Science words
research
coir
cotton
rubber

Natural materials

Think like a scientist!

Sometimes people take a natural material and **change** it. The changed material is more **useful**. They may change the shape of the material. They may wash and dry it. They may colour it.

Wool is a natural material. It comes from a sheep.

We wash it. We spin it into yarn. We change its colour. It is still wool.

We use the wool to knit pullovers or to weave woollen cloth.

Science words
change   useful

1

Let's find out how wool is changed to make it more useful.

a Do research. How can we change a sheep's wool into a pullover?

b Share what you find out. Draw a flow diagram to show the different stages. Draw a different picture for each stage.

2

Look at the pictures below.

a What material is each object made of?

b Which materials are natural?

c Which have been changed?

d Which are manufactured?

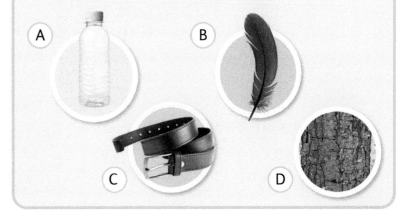

A   B   C   D

Materials from animals

Think like a scientist!

Silk is another natural material. Silkworms are animals that make silk thread. Nothing is added to the silk. People spin the silk thread to make silk cloth. It is not manufactured.

silk cloth

1

This flow diagram shows a model of how natural silk is made into silk cloth.

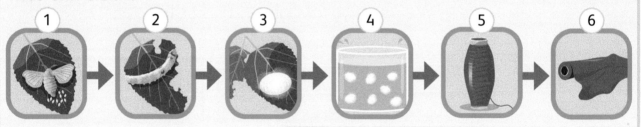

The sentences below are mixed up. Match the sentences to the pictures.

A They spin **cocoons**.

B The eggs hatch and the **larvae** feed on mulberry leaves.

C The cocoons are steamed and washed.

D The cloth is made.

E The threads are wound together.

F A silk moth lays eggs.

2

Look at the sorting circles you made in Activity 1 on page 57.

a You now know more about natural, changed and manufactured materials. Do you need another circle?

b Move any materials that are in the wrong group.

c What made you decide which group to choose for your materials?

Science words

silk

cocoon

larvae

Properties of materials

Think like a scientist!

We can describe materials by how they look and feel. This is called the **texture** of the material. Scientists call texture a **property** of the material.

Different materials have different properties. You may already know about some textures of materials.

This wooden bowl feels smooth.

This rock feels rough.

1

These objects are made of different materials. Use texture words from the list you made in the *Let's talk* below to describe the material. Can you use more than one word?

sandpaper

glass marble

wrapping paper

fabric teddy

Let's talk

What words can you use to describe the texture of a material? Write them in a list. Call your list, *Texture words*.

Challenge yourself!

Ask your family to help you to collect samples of materials. How many different textures can you collect? Bring the different samples to school.

Science words

texture   property

Test materials

Think like a scientist!

Scientists **test** materials. They **investigate** the properties of each material. This is how we learn more about material properties.

We can use our senses to test some properties of materials. We can touch and feel the material. We can try to lift it and try to bend it. Here are some properties to describe materials:

(light) (soft) (heavy) (rigid (stiff)) (hard) (rough)

(smooth) (flexible (bendy))

1

a Choose five objects in the classroom. Test the first object. Find out what properties it has. Is it **rough** or **smooth**? Is it hard or **soft**?

b Write a property of the material on a label. Place it on the object.

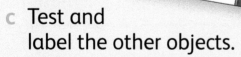

c Test and label the other objects.

d Take photographs of the five objects you labelled.

2

Compare your labels with the labels other learners wrote.

a Did you label each object in the same way?

b Which material has the most properties placed on it?

Science words

investigate   rough

smooth   soft

How many properties?

Think like a scientist!

Materials can have more than one property. A chair needs to be strong and smooth. It also needs to be rigid (stiff). All these properties make a chair useful.

1

A group of learners labelled the objects in their classroom. They put the results in this table, but they made some mistakes.

Object	Material	Soft	Hard	Heavy	Light	Stiff	Bendy	Smooth	Rough
book	paper				✓		✓	✓	✓
desk	wood		✓	✓		✓		✓	
ruler	plastic		✓	✓		✓		✓	
spoon	metal		✓		✓	✓		✓	
jumper	wool	✓			✓		✓		✓
drawer	plastic		✓		✓	✓		✓	

a What are the mistakes with the book/paper, ruler/plastic and jumper/wool?

b Do any materials have only one property?

c Which property did they find the most in their classroom?

2

a Draw a similar table to the one above. Use the photographs you took in Activity 1 on page 62. Tick the properties of the materials you photographed.

b Are you surprised by any properties?

Challenge yourself!

Sort the objects in your photographs into groups. How many different ways can you sort them? Show your partner the different groups you can make.

Describe properties

Think like a scientist!

Scientists share information. Sometimes they use special words. Everyone must agree about what the words mean. Then we can use the word correctly. The meaning of a word is called a **definition**.

The word, bendy, for a wooden ruler is not correct. A wooden ruler is stiff, it does not bend easily. Scientists say that it is **rigid**.

1

Match these property words with their meanings. This will give you a definition of each word. One has been done for you.

(hard) (rough) (smooth) (strong) (**flexible**)

(runny) (**waterproof**) (rigid) (soft)

a A material that is like water. runny

b A material with a surface that is bumpy with ridges.

c A material with an even surface without any bumps.

d A material that is firm and solid, and is not easy to bend, stretch or cut.

e A material that does not allow water to go through it.

f A material that can bend, but will not break.

g A material that is easy to mould, cut, **compress** (squash), stretch or fold.

h A material that is not easily broken.

i A material that does not bend or stretch.

Modelling clay is flexible. We can bend, twist and stretch it.

Science words
definition rigid
flexible   waterproof
compress

Other properties

Think like a scientist!

Science words
transparent   opaque
absorbent

Some scientific words for properties are easy to understand. We might use them when we talk to our friends, such as rough and smooth.

Some words for material properties we may not use every day. We use **transparent** when we talk about a see-through material.

Glass is transparent. This is a property of glass.

The opposite of transparent is **opaque** (say: o-**pay**-k).

Being waterproof is another property of glass. The definition is: does not let water through.

1

Do you know what these words mean?

transparent   opaque

absorbent

Find each definition in the *Science dictionary* at the back of this book.

2

Look at this list of properties and their opposites. There are lots of mistakes! Write the correct list of opposite properties.

hard – rough

strong – heavy

light – shiny

rigid – soft

dull – flexible

Can you add any more property pairs?
Hint: There is already one on this page!

Find the most absorbent kitchen towel

Think like a scientist!

An absorbent material can take in and hold onto water. We say it has soaked up the water.

We can test for the property of absorbency. If a kitchen towel is absorbent, it should hold the water. It will not let water out.

kitchen towel

1

Do all kitchen towels absorb the same amount of water? Let's find out!

a Feel each of the kitchen towels. Describe their texture. Use your observations to predict which kitchen towel will be most absorbent.

b Test your kitchen towels. Follow the instructions in the cartoon.

c Which type of kitchen towel was best? Was your prediction correct?

d Look at the kitchen towels under a magnifier or microscope. How is the most absorbent kitchen towel different to the others?

You will need...
- different types of kitchen towels
- magnifier or microscope
- two rulers
- food colouring
- water
- sticky tape
- tray

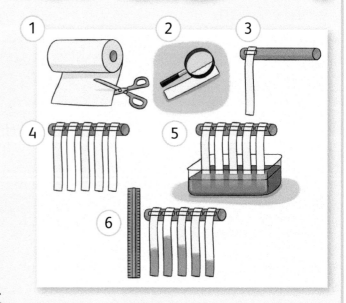

The most absorbent materials: Upsalite

Sometimes scientists forget things. Sometimes they make mistakes. Swedish scientists did both of these and made one of the world's most absorbent materials. They forgot to turn off some equipment in their **laboratory**. They returned to work and found they had made a new material. They had invented upsalite (say: **up**-sal-ite). The scientists observed the material to find out how it worked.

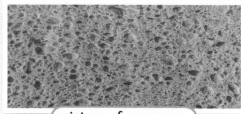

picture of a sponge

Absorbent materials have holes in them. Water runs into these holes. These holes can be very small. We might need a magnifier to see them. Some holes are larger.

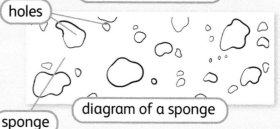

holes

diagram of a sponge

sponge

1

You will need...
- magnifier
- different materials – sponge, fabric, kitchen towel, metal, plastic and paper

a Use a magnifier to look at a sponge. What can you see? Draw a diagram. Remember a diagram does not have colour. It needs labels.

b Now look carefully at the other materials. What is similar about the sponge, fabric and kitchen towel?

c Predict which materials will be absorbent. Sort your materials into two groups.

Science word
laboratory

Sort materials: Venn diagrams

Science words
overlap   ceramic
Venn diagram

Think like a scientist!

Scientists have different ways to share how things are grouped. You can use simple sorting circles to show materials that share properties.

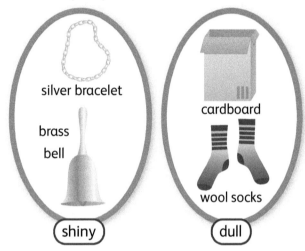

silver bracelet

brass bell

cardboard

wool socks

shiny   dull

Sometimes we use circles that are joined to sort objects. Objects with both properties go where the circles **overlap**.

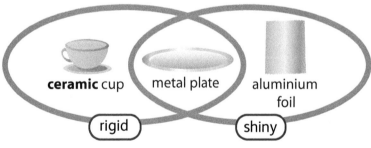

ceramic cup   metal plate   aluminium foil

rigid   shiny

Scientists and mathematicians call these circles a **Venn diagram**.

We do not use a Venn diagram to sort materials that have opposite properties. This is because there would not be any objects in the overlapping part.

1

Play a Venn diagram game. Collect six objects from around the classroom.

a Choose three of the six objects. Think of two properties for these three objects. One object must have both properties.

b Draw a Venn diagram. Sort the three objects into the three sorting circles. Do not label the sorting circles! Make sure you have an object in each circle.

c Show your Venn diagram to a partner. Ask your partner to name the property in each circle. Is your partner correct?

d Try the game again using different objects.

How many uses?

Think like a scientist!

We choose materials for different purposes (**functions**). We use glass for windows. Glass is transparent. We can see through the window. It lets in light. Glass is waterproof. It does not let in rain. A window made of cardboard would not be helpful! A cardboard window will not let in light and will become **soggy** when it rains!

1

We can use materials to make many different objects.

a Choose one of these six materials:

(glass) (leather) (metal) (stone) (paper) (cotton)

b Write a sentence to say what properties the material has.

c Create a **mind map** like the one below. This shows how one thing links to lots of others. Write the name of your material in the middle of your page.

d Show the uses of the material. Write or draw them around the name.

e Share your mind map with a partner. Does your partner agree with your uses for the material?

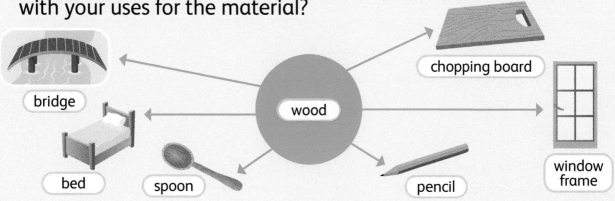

Suitable and unsuitable materials

Think like a scientist!

Materials are chosen for a function because of their properties. Fabric is chosen for clothes. Fabric is **suitable** because it is soft and flexible. We would not find a T-shirt made of stone suitable! It would be very heavy to wear. It would not move when we did. It would be rough and hard.

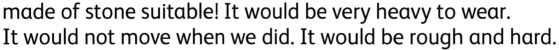

Would we want a saucepan made of chocolate? Chocolate changes shape when it gets warm! We could not cook in a chocolate saucepan. Metal is much more suitable for saucepans.

1

Match these objects to the best materials to make them from. Choose a property from the words below to explain why this material is suitable.

teapot   waste paper bin   drinking cup   toy

wood   glass   metal   plastic   ceramic   fabrics

waterproof   strong   easy to shape   rigid   hard

flexible   soft

Let's talk

What materials would you not use to make each object in Activity 1? Explain why not.

Science word
suitable

Clever and silly choices

Let's talk

Mr Jumble is not very good at choosing the right materials. His shirt is made of candle wax. His shoes are made of brick. Very silly!

Talk about Mr Jumble.

a Why are his clothes not the right materials?

b Can you think of better materials for a shirt and shoes?

c Why are your choices better?

1

a Draw and label a house for Mr Jumble. Choose silly materials for the roof, windows and walls.

b Write a sentence to explain why the materials are not good choices.

c Write a sentence to explain what materials he should use instead.

2

Play this game with a partner.

a Write the names of different objects and materials on two spinners (look at the picture).

b Use a pencil and paper clip. Spin to find an object. Spin to find a material.

c Which are good choices? Why?

d Which choices are silly? Why?

What have you learnt about materials?

1

Are the statements below true or false?

Statements	True	False
a Rigid means that something breaks easily.		
b Absorbent materials let water run off them.		
c Opaque materials stop light from travelling through them.		
d Flexible materials can bend.		
e Shiny materials give off light.		

2

Viti is not sure about the differences between natural, changed and manufactured. Write a note to her to describe the differences.

What can you do?

You have learnt about materials. You can:
- ✔ describe some natural and manufactured materials.
- ✔ describe the differences between natural and manufactured materials.
- ✔ describe the properties of some materials.
- ✔ sort materials into groups and Venn diagrams.
- ✔ explain why a material is used for its purpose.
- ✔ test materials to find their properties.
- ✔ explain what you found out.

What are properties and how can they change?

What do you remember about the properties of materials?

You already know we can change the shape of some materials. We can squash, bend, twist or stretch them. You also know that materials have properties. What properties can you name? Name one material for each property you named.

Think like a scientist!

We can change the shapes of some materials. The material remains the same. It still has the same properties. It is only the shape of the material that changes.

Modelling clay is soft and flexible. These are its properties. If we bend the modelling clay, we change its shape. It is still soft and flexible. It is still modelling clay.

Let's talk

Talk about these materials:

a Which can you bend?

b Which can you squash?

c Which can you twist?

d Which can you stretch?

e What is the name of the material before and after you change it? Has the name changed?

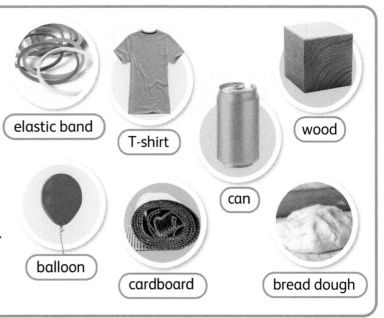

elastic band

T-shirt

wood

can

balloon

cardboard

bread dough

Heat materials

Think like a scientist!

We can change materials in different ways. We can **heat** a material. This may change its shape. It may change the material in other ways.

Let's talk ⭐

Have you ever had chocolate on a hot day? Something happens to the chocolate in the Sun. It changes.

a What properties does cold chocolate have? How does the chocolate change in the Sun?

b What questions could you ask to help you find out what happens? For example, does it still taste like chocolate?

1 ⭐

Work in a small group.

a Talk about what you think will happen to these foods when they are heated.

ice cube

tomato

chocolate

butter

raw egg

b Predict which foods will change when heated.

c Try heating these foods at home with your family to find out.

Science word
heat

Test your predictions

1

Scientists test their **predictions** to see if they are correct.
Let's test your predictions from Activity 1 on page 74.
We can test what happens when materials are heated.

You will need...
- aluminium foil
- selection of foods to test – tomato, chocolate, butter, egg, ice cube
- bowl of hot water
- tongs or a clothes peg to remove the aluminium tray from the water safely

a Make a small foil tray. Put a small amount of the food in the tray. Hold the foil tray with a peg over a bowl of hot water.

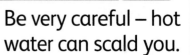
Work safely!
Be very careful – hot water can scald you.

b Write your results in a table like this.

Food	What it looked like at the start	What it looked like when warmed	Changed or not?
butter	yellow block	clear yellow liquid	yes, was runny
bread	white, soft with holes	same	no
ice			
chocolate			

c Were your predictions correct?

d Talk to your group. Look at the materials that changed. Have you made new materials?
Or are the materials just more runny?
How could you test and find out?

Science word
predictions

Cool materials

Think like a scientist!

The **opposite** of heating is **cooling**. When we heat a material, it can change. When we cool a material, it can also change.

Science words
opposite
cooling
frozen
freezer

Let's talk

Have you ever had ice cubes in your water? Ice is very cold water. It has been **frozen**. We put water in the **freezer** to make ice cubes.

a How does ice feel?

b How is this different to water?

1

Work in a small group.

a What do you think will happen to these foods when they are cooled?

b Predict which foods will change when cooled.

tomato   honey   water   milk   butter

c Describe what will happen to each one.

d How can you test your ideas?

Work like a scientist

Scientists test their ideas, even if they think they know the answer. They check to see if they are correct. We can make a model freezer to see if our predictions earlier were correct.

> **You will need...**
> - transparent plastic bags, that seal
> - different foods to test – tomato, butter, water, milk, honey
> - transparent bowl of ice and salt water
> - tongs or a clothes peg to remove the plastic bag from the water safely

a Put a small amount of each food into a plastic bag. Seal the bag. Put the bag into the ice. Leave it for ten minutes. Take it out. Feel the food through the bag. Observe any changes.

Work safely! ⚠ Be very careful – ice can make your fingers numb.

b Write your results in a table like this.

Food	What it looked like at the start	What it looked like when cooled	Changed or not?
butter	yellow block, can be squashed	very hard, can be broken	yes a little
water	runny and transparent	hard, transparent	yes

c Were your predictions correct?

d Talk to your group. Look at the materials that changed. Have you made new materials? Or are the materials just harder? How could you get back the materials you started with?

Make new materials

Think like a scientist!

You have found out that heating can change materials. Cooling can also change materials. Many of the materials can also be changed back again. If we heat chocolate, it becomes runny. If we cool chocolate, it goes hard again.

When we heat some things, we cannot get them back to the way they were. An example of this is cooking food.

When we heat an egg, the egg changes. We cannot change the egg back to a raw (uncooked) egg. Cooking is a scientific activity!

1

We can make new materials that are not food. This potter is making a pot from clay. When soft clay is heated it changes. The potter is using science just like you.

You will need...
- modelling clay that can be heated in a cooker
- cooker

a Describe the properties of the clay before you model it.

b Predict what will happen to the clay when you heat it.

c Make your own clay pot. Bake it in the cooker.

d Take the pot out of the cooker. Describe how the clay has changed. Was your prediction correct?

Work safely! ⚠️

Do not touch your pot until it has cooled.

Let's cook!

1

This is Louis. He is a scientist. Can you work out what kind? **Hint**: These scientists make new materials with food. Chefs mix food and heat it. This changes it into something delicious to eat.

Be a chef scientist! Make chocolate crispy cakes. Use this recipe.

You will need...
- 225 g chocolate
- 125 g cereal (puffed rice or corn flakes)
- three tablespoons of golden syrup
- handful of dried fruit
- kitchen scale or digital scale
- mixing bowl and spoon
- cooker or microwave oven
- paper cake cases

What to do:

1 Use a scale to measure your **ingredients**.

2 Ask an adult to help you to heat the chocolate until it is runny.

3 Add the syrup and cereal. Stir until it is all covered in chocolate.

4 Add dried fruit, if you like. Stir again.

5 Put spoonfuls of the mixture into the paper cake cases.

6 Let the cakes cool until the chocolate is hard.

Your cakes are ready to eat! How do they taste?

Science word
ingredients

Challenge yourself!
Try this recipe at home with your family.

Science in context

Invent new materials

Some scientists are called **inventors**.

Sometimes inventors **invent** new materials. They change the materials they have and make new, different materials.

Sometimes inventors use an existing material in a different way or for a different job.

Inventors think about things and try new ideas.

John Boyd Dunlop was a Scottish inventor. He lived over 150 years ago. He noticed his child was uncomfortable riding their tricycle. It had three metal wheels. He realised that rubber wheels would be more comfortable. He then decided to put air inside the rubber tyres. This made them bouncy. It also meant they could travel further before wearing out. This is how John Boyd Dunlop invented the pneumatic (say: new-**ma**-tick) tyre.

Science words
inventors
invent

John Boyd Dunlop

1

Read the information about the inventor John Boyd Dunlop. Use the information to complete a fact card about him. Include:

- Date of invention
- Name of invention
- Uses of invention

Find a picture of John Boyd Dunlop and stick it on your fact card.

2

Research another inventor. You could choose Spencer Silver, Ruth Benerito or Charles Macintosh, or you could choose another inventor. Create a fact card them.

Display your fact cards in your classroom.

What have you learnt about changes to materials?

1

Are these statements true or false?

Statement	True	False
a Heating a material always makes a new material.		
b Cooling a material can change its properties.		
c Heating a material can make it runny.		
d We can sometimes change a material back that has been heated.		

2

Preti the plastic penguin is trapped in the ice. Think about how materials can change. What is the quickest way to get Preti out without damaging it? Why?

What can you do?

You have learnt about changing materials. You can:
- ✔ recognise that some changes make new materials.
- ✔ describe how some materials can be changed.
- ✔ describe how we can make a new material.
- ✔ describe how people have made new materials over many years.
- ✔ make and record observations in tables, and describe what happened.
- ✔ use equipment safely.

1 Look at the objects.
  They are made from
  different materials.

  a Name one that will bend.

  b Name one that will stretch.

  c Name one that will squash.

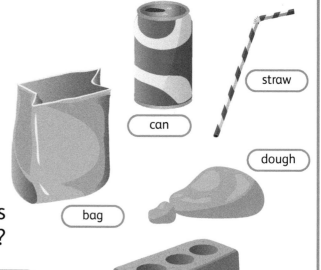

straw

can

dough

bag

2 Glass is a material. Which words
  describe the properties of glass?

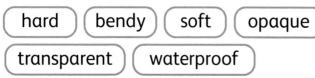

  hard   bendy   soft   opaque

  transparent   waterproof

brick

3 Rock is a natural material. Name two other natural materials.

4 Name two ways we can use wood.

5 a Wool is a changed natural material. Name another
     material that we change to make it more useful.

  b Describe the difference between a manufactured and
     a changed natural material.

6 Which of these materials are manufactured?

  a nylon   b clay

  c wool   d plastic

  e metal

7 Plastic is often used to make toys for babies. Name two
  properties that make it a good material to use.

8   Copy the sentences. Fill in the gaps with the correct words.

   | test | | property | | materials | | experiments |

   Scientists carry out lots of _____. They _____
   materials to find out their properties. _____ can have
   more than one _____.

9   a   Choose which materials these objects can be made from.
        Use these words:

   | plastic | | glass | | leather | | wood | | gold | | rubber | | cotton |

   A   B   C   D

   E   F   G

   b   Choose one of the objects above. Explain what properties
       that material has that makes it suitable for the object.

10 Describe what must be done to water to change it to ice.
    Choose the correct word.

   | cool | | freeze | | heat | | irreversible | | melt |

   Water needs to _____ to turn it to ice.

11 Name two jobs people do that use the science of materials.

Pushes and pulls

What do you remember about forces?

You have used forces before. They can change the shape of materials. Which forces do you remember? We can use words to show our actions (movements).

Pull

1 Write the other words you know for forces to show the action.

2

a What is happening in each picture? Is there a push, a pull, or both?

b Sort them into groups. Use a Venn diagram like this one.

push pull

What have you pushed and pulled today?

Mime forces

Science words
force   push   pull

Think like a scientist!

A **force** is either a **push** or a **pull**. Pushes and pulls make objects move. For a push or a pull to happen there must be two objects. Often you are one object. What you are trying to move is the other object.

A push force moves something away from you.

A pull force moves, or brings, something towards you.

1

We can mime forces by only using actions and no words.

Take turns with a partner.

Mime putting on a piece of clothing. Ask your partner what piece of clothing it was. Did you use a push or a pull to put it on?

2

Look around your classroom. Draw a picture of it. Use one colour to label the things you need to push. Use a different colour to label the things you need to pull. Do you use more pushes or pulls in your classroom?

Challenge yourself!

Sometimes we must use a push and a pull to get dressed. Try putting on your socks without pushing your feet into them! What other clothes do you need a push and a pull for? Try it at home!

Change the shape of objects

Think like a scientist!

Forces can make some materials change shape. We can use a pushing or pulling force. Some materials stay in their new, changed shape even when we stop pushing or pulling them. An example of this is butter.

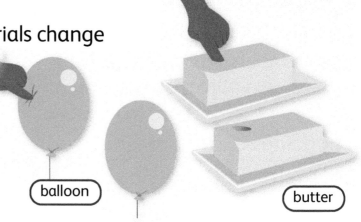

balloon

butter

Some materials go back to the shape they were when the force is removed. Look at the balloon example in the picture.

1

Let's model how forces change the shape of an object.

You will need...
- elastic band
- modelling clay

a How can you change the shape of an elastic band? Can you pull it? Can you push it?

b How many ways did the force change the elastic band? Look carefully at the size, colour and shape.

c Does the elastic band stay in the shape when the force is removed?

d Try the same activity with modelling clay. What happens when you pull or push the modelling clay? What happens when you stop pushing or pulling it?

e Which material stays the way the force has changed it? Which material goes back to the shape it was?

Modelling clay game

1

You will need...
- modelling clay
- numbers 1 to 6 written on small squares of paper (in a bag)

Games can help us explore how different forces change the shape of an object. Scientists learn from playing games.

a Play a game.

- Start with a small ball of modelling clay.

- Choose a number from the bag. Look at it. Then put it back into the bag.

- Each number is a different action.

  Key:
1 – pull	3 – twist	5 – squash
2 – push	4 – roll	6 – stretch

- Change the shape of the modelling clay using the action. For example, if you chose 4, then roll the modelling clay in your hands.

- Take turns. Have at least 20 turns each.

b If you choose 3, 4, 5 or 6, describe the forces you are using.

Remember, all forces are either a push or a pull, or both, like a twist!

Challenge yourself!

Play the game again. This time, try to make an animal, such as a cat or a bird. Use the action numbers you take out of the bag. Who made their animal first?

Let's get moving!

Science word
effect

Think like a scientist!

Objects do not move by themselves. They need a force to make them move. We cannot see a force. We can only see the **effect** the force has on an object. We see how it moves or how it changes. Objects also need a force to stop moving.

1

You will need...
- marbles or table tennis balls
- pieces of paper
- straws

a Put a marble on a desk or on the ground. Make sure it is still. Ask your partner how you can make the marble move. Will it move if you do not touch it?

b Try out your ideas. What happened?

c Talk about how you can make your marble move in a different way. Try out your ideas.

d Once your marble is moving, predict how you can stop it. Try out your ideas. Which worked best?

e List all the ways you can make the marble move. What force are you using? List the ways you can stop it moving.

How far?

Think like a scientist!

Scientists describe a pattern between the size of the force and how far an object moves. We can discover the pattern with a simple test.

1

You will need...
- marbles
- something to measure distance, such as fingers, string, straws or blocks

a Put a marble on the floor or desk. Make sure it is still.

b Use a small push to move the marble. Measure how far the marble moves.

c What do you think will happen with a bigger push? Try it. Measure the distance the marble moves.

d Try pushing the marble with a medium push. Measure how far the marble moves. Copy this table. Write in your results.

Size of push	Distance moved
small	
big	
medium	

e Describe what happens to the distance the marble moves as you use bigger forces. This is a pattern. Use these words:

bigger   further   smaller   shorter

f Predict what happens to the distance the marble moves if you use a very, very big push. Try it!

89

Change direction

Think like a scientist!

Forces are very important. Without a push or a pull nothing will happen. Forces are needed to make objects move and change shape. They are also needed to make objects change **direction**. Backwards and forwards are two different directions. So are left and right. There are two more directions. What are they?

Science word
direction

Let's talk

Look at these children. Which force (push or pull) would they use to make the object change direction?

1

Look at the above pictures.

Mime playing a sport with your partner. Use a push or a pull to make an object change direction.

Try miming tennis, swimming, football or cricket.

How fast?

Think like a scientist!

Scientists know that pushes and pulls can change how fast an object moves. There is a pattern linking the size of the force and how fast an object moves. Scientists find patterns that help them to make predictions.

1

Some learners tested the size of a push and how fast a ball moved. They used different sized pushes. They counted how long it took the ball to reach the wall. They counted, one elephant, two elephants. Saying 'one elephant' takes approximately one second. These are their results:

a Sort the results by size of push.

b Copy the table. Write the results into your table in the correct order.

c Copy the blank graph. Draw a block graph of the learners' results.

d How long did it take the ball with a very small push to reach the wall?

e What is the pattern between the size of push and how long it takes the ball to reach the wall?

f Try this activity with a partner or at home. What pattern did you get?

Size of push	Elephant counts
very small	6
very big	2
small	4
big	3

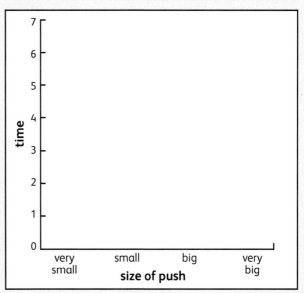

Squirty forces

Let's talk

a What is Maris doing to make the water squirt out of the bottle?

b What does the water squirt do to the skittle?

1

We can make a model or game to show how forces work.

You will need...

• six plastic water bottles with a small amount of sand or pebbles in each one (as skittles)

• squeezy bottle for squirting

• water

What must you do to push more skittles over at the same time?

a Make your own squirty game.

b What will you do to make the water knock down the skittles?

c How can you make the skittles fall in different directions?

2

a Use your squirty water bottle to make a ball move in a straight line across the playground. Where do you squirt the water?

b Now make the ball zig-zag. What is different about how you squirt the water?

c Predict what happens if you squeeze the bottle with more force.

d Test your prediction. What other tests can you try?

Blow football

Science word
blowing

Think like a scientist!

You can make air move by **blowing**. When you blow, you push air out of your mouth. This air can push some objects. You tried making marbles move in this way in the activity on page 88.

1

You will need...
- two table tennis balls
- two straws

a Sit on the floor next to a partner. Place the balls in front of you.

b Each blow through your straws to try to make the balls move.

c Which ball moved the furthest? Why? How could you tell?

d Predict how you can make the ball move even further. Test your idea.

Let's talk

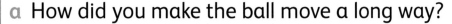

Talk about what you did in Activity 1. Use these words to answer the questions below:

straw   ball   push   change   direction   air

a How did you make the ball move a long way?

b How did you make the ball move a little way?

c How did you make the ball go faster?

d How did you make the ball change direction?

e How do you stop the ball?

I blew hard to make the ball go faster. The air pushed the ball and it moved faster.

What is your question?

Think like a scientist!

Scientists ask lots of questions. Then they test and do experiments to find answers.

These learners asked some questions about making things move.

One question was: How many different things can I move by blowing through a straw?

1

a Answer your questions from the *Let's talk* activity.

b If you have a camera, take photographs of what you decided to do.

c Share what you did to answer your questions. Show another pair of learners what you did.

d Look back at your prediction. Was it correct?

e What was the pattern in your investigation? Describe it.

Let's talk

a Think of questions to investigate how we can make things move. Think about moving them by blowing through a straw. Use some of these words in your questions:

small flat

round big

big force

small force

changing direction

b Write down your questions. Choose one to investigate.

c Predict what the answer will be.

d Decide what you will do to answer your question.

Science in context

A day in the life of a tennis player

Scientists are not the only people to use forces. We use forces in everyday life. We use them to get dressed and to cut our food.

We use forces to play sports. Footballers, cricketers and tennis players all change the direction of a ball. They use the science of forces to do this.

Sergio is a tennis player. He practises every day. Look at the pictures of how he spends his morning. Find all the forces he uses.

6:30 am

7:00 am

7:30 am

9:00 am

9:15 am

10:30 am

11:00 am

Challenge yourself!

Watch people at home. What forces can you see them using?

What have you learnt about forces and energy?

1

Are these statements true or false?

Forces can:	True	False
a make things start moving		
b make things grow		
c make things eat		
d make things change shape		
e make things stop moving		
f make things think		
g make things speed up		
h make things change direction		
i make things get smaller		

What can you do?

You have learnt about forces and energy. You can:
- ✔ name the two forces that make things move.
- ✔ say how a force starts the movement of an object.
- ✔ say how forces change the movement of an object.
- ✔ say how forces can change the shape of an object.
- ✔ describe a pattern linking a force and the movement of an object.
- ✔ use games to model what forces do.
- ✔ ask questions and describe what happens.
- ✔ present results in graphs and tables.

Light to see

What do you remember about light?

Do you know what we use to see? **Light** is the other thing we need to be able to see. Draw a picture of what it would be like if you could not see.

Science words
light
dark

Let's talk

a Look around. Is it daytime? Is it light or **dark** outside? Are lights on in the classroom? How can you see things? Which sense do you use?

b Look at the picture. How does the room look different in the dark?

1

Work in a small group.

a Talk about light and dark places.

b Make a table of light and dark places like the one below:

Light places	Dark places
outside in the day	in a cinema
in my classroom	under the bed

c Look at your list of dark places. How can you make them light?

d Are the dark places completely dark?

Let's talk

Talk about when you have been in the dark.

- How did it feel?
- Did you like it?
- Do you prefer being in the light? Why?

Too many lights?

Think like a scientist!

Darkness is the **absence** of light. The absence of something means it is missing. Light is missing.

So many lights are left on in towns and cities that it is never completely dark. Lights burn brightly all night.

1

This photograph shows the Earth at night – from **space**. Look at all the lights!

a   Where are the most lights? Why do you think this is so?

b   Where are the **fewest** lights?

c   Do you see a difference between land and water? Why are there no lights in the oceans? Try to use this scientific word – absence.

Science words
absence   space   fewest

Make a dark den

Think like a scientist!

It can be difficult to find a dark place. In a dark place there is no light. We cannot see if there is no light.

1

You will need...
- table
- large blanket, curtain or thick cloth

a Make a dark den with your partner.

- Find an unused table that is big enough for you to sit under.

- Cover the table with a large blanket, curtain or thick cloth. The fabric must touch the floor.

- Make sure there are no gaps where the light can get in.

b Sit in your den. Is any light coming in? If you see strips of light, add more fabric to cover the gaps.

c Sit in the dark. What can you see? Can you see your hand in front of you?

d What can you use to help you to see in your dark den?

e Hide something in the den for your partner to try to find. How easy is it to find?

Sources of light

Science words
source
Sun

Think like a scientist!

Scientists say that light comes from a source. The **source** of something is where it starts.

A light source makes its own light. There are many different light sources.

flashlight   fire   lamp

The **Sun** is the most important source of light.

the Sun, shining in the sky

Let's talk

Light comes from many sources. Think about all the things that make light. Write them in a list.

1

a Look at these pictures. Which objects are sources of light? Sort them into two groups. Can you use sorting circles?

match   flashlight   TV

mirror   candle   ring

b Discuss the objects that are not light sources. Why are they not light sources?

Which objects would make light in your dark den? Which could you see in the dark?

Brightness

Science words
brighter
dim

Think like a scientist!

Some light sources are **brighter** than others.

Very bright light sources include the Sun and floodlights in a sports stadium.

Some sources of light are not as bright. Scientists say they are **dim**. A nightlight has a dim light.

1

These pictures show sources of light. Put them in order from brightest to dimmest.

A floodlight

B small flashlight

C Sun

D car headlight

E candle

F kitchen light

2

You will need...
- two flashlights

a Turn on the two flashlights. Which is the brightest? Which is the dimmest?

b Predict the best flashlight to help find anything you left in the den from page 99.

c Use the flashlights in your dark den. Which is brighter?

Dim or bright?

Think like a scientist!

Some sources of light are dim. They are not very bright.

Matches, candles or small flashlights may give dim light.

Lamps are sources of light that are used in many homes.

Some lamps give a dimmer light than others.

Let's talk

List the light sources in your school.

a Which is the dimmest? Which is the brightest?

b When would you need a very bright light?

c When would you want a light to be dim?

1

Play a game of charades – an action game with no words.

a Choose a light source. Mime it to your group. Use actions but no words to describe it.

b Can the rest of the group guess the light source?

Did you know?

The world's brightest flashlight can light objects that are up to 14 kilometres away. This flashlight is as bright as 52 million candles!

Science in context

Is bright light good or bad?

The Sun is our brightest light source. It can harm us. On sunny days, we need to wear a hat and a T-shirt. We need to protect our bodies when we are in the Sun. We can stay in the shade or use suntan lotion. This will protect us from getting sunburnt.

A lighthouse has a very bright light. At night we cannot see the Sun shining. We cannot see easily. When it is foggy it is not easy to see. People used to light big fires on top of cliffs. Then they started to build a tower for the fire.

Lighthouses today are still built in tall towers, but now we have electric lights.

1

Find out how to stay safe in the Sun. Make instructions for younger learners. Make a poster, or a leaflet, or do a role play to show them what to do. Share this with younger learners.

2

Describe how lighthouses have changed over time. Find out more from a book or the internet. Share what you find out with your class.

What have you learnt about light?

1

Mrs Ramirez has a dark classroom. It is hard to see all her learners. She has a lamp on her desk. It does not help her to see the learners very well.

a How can Mrs Ramirez make the classroom brighter?

b Draw a diagram with labels to show what Mrs Ramirez needs to do.

What can you do?

You have learnt about light. You can:
- ✔ name some sources of light.
- ✔ name the brightest light we have.
- ✔ describe what darkness is.
- ✔ describe what it is like to be in the dark.
- ✔ know if we can see in the dark.
- ✔ sort light sources by how bright they are.
- ✔ do research into how to stay safe from our brightest light source.

What does electricity do?

What do you remember about electricity?

Electricity makes things work.

a Sort these objects into groups by what the electricity makes them do.

TV | TV remote control | flashlight | mobile phone | kettle

b Write a list of objects in your classroom that use electricity.

c Add the items from your list to the groups. Are there any objects that do more than one thing?

Think like a scientist!

Scientists have a special word for objects that use electricity. Do you remember what this is? We call them **appliances** (say: a-**ply**-an-sez). Practise saying this word to a partner.

We can easily tell if an appliance uses mains electricity. It has a **plug**. We put the plug into a **socket**.

wall sockets

plug | plug | plug

Let's talk

What will happen if we do not plug in an appliance?

Science words
appliance   plug   socket

Science in context

Life in great-grandma's day

We have not always had electricity. Talk to some older people about what life was like in the past. Electricity makes everyday tasks easier. My great-grandma explains how it was in her day:

'When I was your age, I walked to school every day. We did not have motor cars or motorbikes. We walked home in the dark in winter. We did not have flashlights. My mother spent all day doing the washing in a big tub. She heated the water over a fire. Then she would hang the washing on a line to dry. She also used the fire to help cook dinner. We did not have a cooker.

In the summer, the weather was beautiful, but very hot. We did not often have ice cream. We did not have a freezer, but my friend did. We did not have air conditioning. We made fans out of paper and waved them in front of us to keep cool.

It was not all bad though! We did not have mobile phones. We did not have homework. We played games with our family instead of watching television. That was fun.'

Imagine a world without electricity.

a What would you miss most? Why?

b Do you think you would like it if we did not have any electricity? Why?

c What questions would you ask someone who has never used electricity? What would you tell them about appliances?

Electricity from a cell

Think like a scientist!

You already know that some appliances use mains electricity. Some appliances in our homes and school use **cells** (batteries) to make them work. Cells come in all shapes and sizes.

Work safely! ⚠️

Never cut open a cell. It can be very dangerous!

Science word
cell

Let's talk

Talk about the things you use that need cells to work.

a Make a list.

b Share your ideas with other learners.

What appliances could we use during a power cut? That is when there is no mains electricity.

1 ⭐

a Make a list of appliances. Think about appliances from home and school.

b Record which appliances use mains electricity, and which use cells. Use a table like this one:

Appliance	Uses mains electricity	Uses cells
lamp	✓	
mobile phone		✓
radio	✓	✓

c What other appliances use either mains electricity or cells? Look around your home for more appliances.

Let's talk

What is similar about all the appliances that use cells?

Stay safe around electricity

Think like a scientist!

Electricity is very powerful. It helps us do many useful things.

However, electricity can be very dangerous. We need to use it carefully. We need to use it correctly. Then we will be safe.

Let's talk

Look carefully at this picture.

a How many dangers can you see?

b Write down your ideas.

1

a Find out about the dangers of electricity. These questions may help you:

  • What happens if water gets near electricity?

  • What can happen if electrical cables are left along the ground?

  • How can electricity cause fires?

b Make a poster to show and explain some dangers of electricity.

c Now that you have done more research into electricity, try to find more dangers in the picture above.

Appliances have changed our lives

Think like a scientist!

You know that electricity can come from the mains or a cell. You also know that we can use electricity to make things move, light up, heat up, or make sounds. Electricity is very useful.

1

a Read page 106. Use the information from great-grandma to name appliances that have changed our lives. Make a table like this one:

What we used to do	What we do now
use a broom	use a vacuum cleaner
wash clothes in a tub	

b Think of other things that we would have to do if we did not have electricity. For example, how could we clean carpets or have lights at night? Add them to the table.

Science words
robot   gadget

2

Design an electrical machine or a **robot** – a human-like machine. Design it to help you to do work around your home.

a Draw a picture of your robot.

b Include as many **gadgets** (tools) as you can.

c Write labels to show what your robot can do and how it works.

d Share your design with other learners.

e Which is the best design? Why?

Make electrical circuits

Science word
circuit

Think like a scientist!

Scientists often make models. A model helps us understand how something works. We can make models of electrical **circuits**. A circuit is a path that electricity takes.

A circuit needs a source of electricity. It can come from mains or cells. The electricity must travel from a source, all the way around a circuit, to an appliance.

1

Make a circuit. All the parts of the circuit must be joined together.

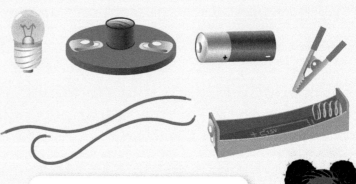

You will need...
- lamp
- wires
- cell
- cell holder
- lamp holder
- crocodile clips

a Use the equipment to light a lamp.

b Draw the circuit you have made to light the lamp.

c Label the things you used.

How many different ways can you find to light the lamp?

Does it matter how you join all the things in the circuit?

What is the least number of parts you could use?

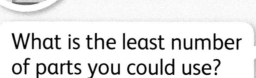

Explore cells

Think like a scientist!

To make a circuit work we need to join everything up correctly. Scientists say the circuit must be **complete**.

Special scientists called **electricians** (say: el-ec-**trish**-ans) use **symbols** to help join all the parts of a circuit correctly. A symbol is a shape or sign that tells us something. We need to observe a cell carefully to find the symbols.

1

You will need...
- cell

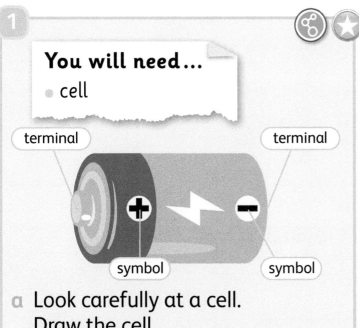

terminal   terminal

symbol   symbol

a Look carefully at a cell. Draw the cell.

b Each end is called a **terminal**. Do you see the symbol at each terminal? Write the symbol on your drawing. Make sure it is written on the correct terminal!

c Share your drawing with a partner. Have you drawn a diagram or a picture?

2

You will need...
- lots of different cells

a Look carefully at each cell. Draw the cells and find the terminals. Write the symbols on your drawings.

b What do you notice about each cell? Share your ideas with a partner.

Science words

complete	electrician
symbol	terminal

Explore lamps

Think like a scientist!

The parts that make up a circuit are called **components**. A cell is a component. So is a lamp.

We make an electrical circuit by connecting components. The leads that connect the cell to the lamp are also components. These make a path for electricity. The path must not have any gaps in it.

1

You will need...
- glass lamp
- magnifier

a Use the magnifier to look at a glass lamp. Draw what you see.

b Look for the wire running through the inside of the lamp.

c Look closely at the outside of the lamp. Find where the wire inside is attached to the outside of the lamp.

d Can you see any gaps in the wire inside the lamp? Why do you think this is?

Electricity moves through the lamp and makes the **filament** wires glow.

Science words
component
filament

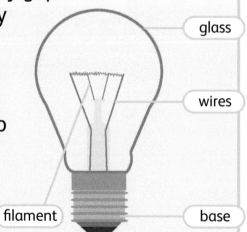

glass

wires

filament

base

Let's talk

Talk about the two drawings of the lamp on this page. How can you tell which one is a diagram?

Work safely!

Never squeeze a lamp. The glass is thin and could break.

Explore circuits

Think like a scientist!

When a lamp lights up, scientists describe the electricity as flowing. When something flows, it moves easily.

Electricity flows from the cell, through one wire to the terminal of the lamp. It flows through the lamp. Electricity flows back along another wire from the lamp to the other terminal on the cell. This makes a complete circuit for the electricity.

1

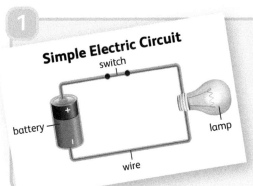

Simple Electric Circuit

Look at the circuit in the picture.

a Can you use your finger to show how electricity flows around the circuit?

b Do you think this lamp will light up? Explain why to a partner.

2

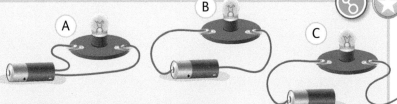

You will need...
- circuit components

Learners made circuits to model electricity working. They drew each circuit they made.

Work in a group.

a Predict which circuits will not make the lamp light up.

b Explain to each other what is needed to make a circuit work.

c Make this circuit.

d Was your prediction correct?

e What rule would you share with the learners who made these circuits so that all their circuits light a lamp?

Make circuits work

Think like a scientist!

Some circuits may have more than one cell in them. We must join the cells up correctly. The terminals must be joined in a certain order. The **positive** (+) terminal in the first cell must touch the **negative** (−) terminal in the second cell.

Science words
positive (+)
negative (−)

1

You will need...
- circuit components

Below are drawings of circuits that do not work.

a Electricians make predictions. Predict what you need to do to make these circuits work.

b Make these circuits yourself. Then use your ideas.

c Were you correct?

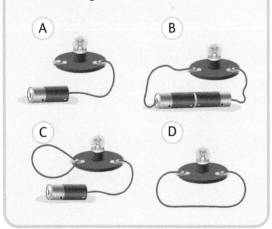

A   B

C   D

2

Electricians also write instructions to help us make circuits work. Write a checklist for other learners to follow if their circuit does not work.

3

Play this spinner game.

a Start with one cell.

b Take turns to spin and collect the component you land on.

c Who can make a complete circuit?

d How many components did you need to collect?

How does a flashlight work?

Think like a scientist!

A flashlight is an appliance that uses electricity to give us light.

In Class 1, you took apart a flashlight to find out what was inside. You know that a flashlight uses cells as a source of electricity. A flashlight often has more than one cell in it. This can make it brighter.

1

You will need...
- flashlight
- cells

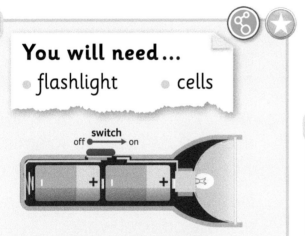

Find out how a flashlight works.

a An adult will help you to carefully take apart a flashlight.

b What is the job of each part? How does each part make the flashlight work?

c Draw a diagram to show how the flashlight works. Label each part and write what it does. For example:
Switch – switches flashlight on and off.

2

You will need...
- flashlight
- cells

a Look closely at the inside of the flashlight where the cells go. What do you notice?

b Can you see any symbols? What do these mean?

c Put your cells into the flashlight. Does it work? Put the cells in differently. Does it work? Why?

d How do the symbols on the cells and on the flashlight help us put it together correctly?

Investigate batteries

Science word
battery

Think like a scientist!

The scientific word for two or more cells in a circuit is a **battery**. We must put cells into electrical appliances in the correct way. We look for the symbols in the cell holder of the appliance.

1

a Look carefully at this selection of cells. They are different sizes and shapes.

b Match the cells to the appliances below.

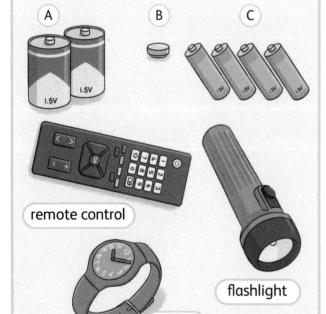

remote control

flashlight

watch

Let's talk

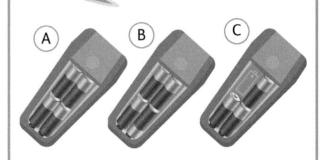

a Look at the pictures of television remote controls. Predict which ones will work.

b Which remote controls will not work. Why? Use science words.

c Describe how to make each remote control work.

d Look carefully at remote control C. What information tells you which way round to put the cell?

e Share your ideas with other learners. Do they have the same ideas?

Challenge yourself!

Write a set of instructions for putting cells into an appliance so that it works.

Broken circuits

Think like a scientist!

A circuit needs to be complete for electricity to flow. A break (gap) in a circuit means that electricity cannot move through the circuit. The component will not work.

1

You will need...
- balloon

Work in a big group for this model of a circuit.

a Stand in a circle. This is like an electrical circuit.

b Pass the balloon to the person next to you in the circle. Do not use your hands! The balloon models electricity flowing around the circuit.

c Your teacher will move two or three learners out of the circle to make a gap. This is a break in the circuit. The electricity cannot get to the other side.

d How can you make electricity flow again?

2

You will need...
- drawing materials
- circuit components

a Take turns to draw a picture of a circuit with a lamp.

b Challenge your partner to explain if the circuit will light the lamp or not.

c Check your ideas by making the circuits.

d Were you correct?

What have you learnt about electricity?

1

Which of these statements is true about staying safe with electricity? Which is false?

Statement	True	False
a Leave cables trailing along the ground.		
b Do not put anything into a socket except a plug.		
c Water and electricity mix well.		
d Only put one appliance in each socket.		

2

a Which of these circuits will work?

b How can you change the circuits that do not work, so that the lamp lights up?

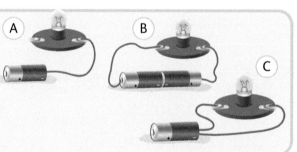

What can you do?

You have learnt about electricity. You can:

✔ name some uses of electricity.

✔ describe the dangers of electricity and how to use it safely.

✔ name some components of a circuit.

✔ describe the path of electricity in a circuit.

✔ make a simple circuit with a lamp.

✔ explain why some circuits work and some do not.

✔ make predictions about which circuits will work.

✔ follow instructions for practical work to stay safe.

1 Copy and complete the sentence. Choose the correct word.

[loss]   [absence]   [brightness]

Where there is no light, it is dark. We say that darkness is the _____ of light.

2 The lamp in this circuit will not light. Why not?

3 a Write the name of the two types of forces.
  b Name one object that can be moved with each force.

4 Which of these circuits will work?

5 Which of these are sources of light?

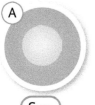
Sun

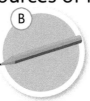
pencil

lit match

mirror

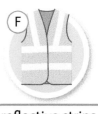
reflective strips

flashlight

6 Write the names of two appliances that use mains electricity.

7   Jack was playing football. He wanted to score a goal. He had to make the ball go further and faster. What should he do to the ball? Complete the sentence:

Jack should kick the ball _____.

8   Viti made a circuit. The circuit did not work. The lamp did not light. What must Viti do to make the circuit work?

9   Write two things that use cell electricity.

10 Can you see anything when it is dark? Explain, using the word absence.

11 Which picture shows that the learner is:

| changing direction | speeding up | slowing down |

Let's rock!

What do you remember about planet Earth?

You may already know what the Earth is made of. Talk about what you know about our planet Earth. These questions may help you:

a What is the Earth mostly covered in?

b What are the other parts of the Earth made of?

c What lives on our planet?

1

Many materials that we use are **extracted** from planet Earth. They have different properties. A group of learners made a list of properties of materials. They have made some mistakes.

a Which of these are not properties?

| colour | texture | hardness |

| use | absorbency | cost |

| name | shape | pattern |

b Can you think of any properties that the learners missed out?

Think like a scientist!

You will find out about a natural material in this unit. We can find this material underground. We can dig to find it. Sometimes it is just lying at our feet. This material is rock.

Science word
extracted

Describe rocks

Think like a scientist!

There are lots of different types of scientists. They have different names. Scientists who study rocks are called **geologists** (say: **gee**-o-lo-jists). They sort and identify or name rocks. Geologists need to be good at observing.

1

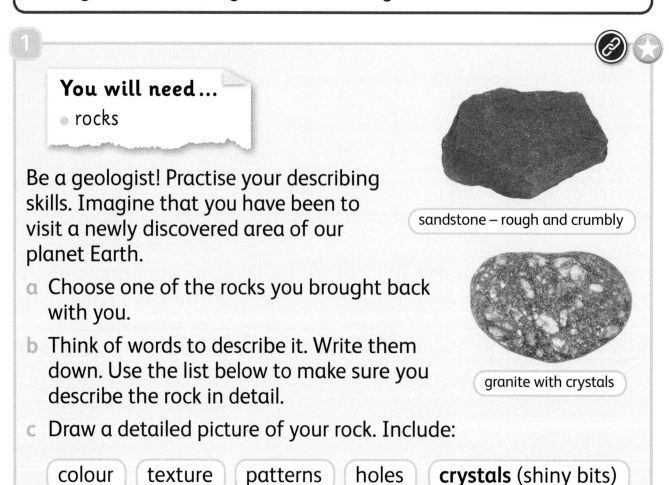

You will need...
- rocks

Be a geologist! Practise your describing skills. Imagine that you have been to visit a newly discovered area of our planet Earth.

sandstone – rough and crumbly

granite with crystals

a Choose one of the rocks you brought back with you.

b Think of words to describe it. Write them down. Use the list below to make sure you describe the rock in detail.

c Draw a detailed picture of your rock. Include:

colour   texture   patterns   holes   **crystals** (shiny bits)

Let's talk

Share your description and drawing of your rock. Can you guess which rock your partner was describing?

Science words
geologist
crystal

Science rocks!

Think like a scientist!

Geologists are good at observing. They are also good at describing things in detail. They look at lots of properties of rocks and describe them. You have looked at properties of materials before.

Let's talk

Take turns.

a Think of one of the rocks you sorted in Activity 1. Ask your partner to guess which rock you are thinking of.

b Your partner can ask you questions about the rock. You may only answer yes or no.

c Which are the best questions to ask?

1 ⭐

You will need...
- rocks
- magnifier

Be a geologist again!

a Geologists must find rocks to study. Go outside and find rocks to look at.

b Geologists observe rocks. Study the colour and texture of the **surfaces** of the rocks. Are they rough or smooth?

c Look closely through a magnifier. Are the rocks the same all the way round?

d Geologists group their rocks. Group your rocks. Use sorting circles or a table. What name will you give each group?

Science word
surface

Test rocks

Think like a scientist!

All materials, including rocks, have properties. As scientists, we test materials to find their properties. Geologists test rocks for properties, such as how hard they are. Hardness is a property of a material.

Let's talk

Talk in your group about how to find out which rock is the hardest.

What evidence can we collect?

How many rocks should we use?

How do we record what we find?

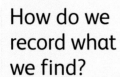

1

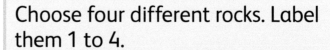

You will need...
- rocks
- magnifier
- cloth

Choose four different rocks. Label them 1 to 4.

a Scratch each rock with your fingernail. Can you see the scratch marks?

b Scratch each rock with a coin. Can you see the scratch marks?

c Scratch each rock with every other rock.

d Make a prediction: which rock is the hardest and which is the softest?

e Rub your scratches on the rocks with a cloth. Sometimes the marks disappear!

How will you make sure you scratch each rock the same way? Scientists try to test materials in the same way.

Look at results

Science word
scale of
hardness

Think like a scientist!

A hard rock marks a softer rock by scratching it.
Friedrich Mohs was a German geologist. He invented a **scale of hardness**. We call this the Mohs scale. He carried out tests like the one you did on page 124. The scale shows that a diamond is the hardest rock. No other rock can scratch it. Talc is the softest rock. All other rocks can scratch it.

1

A group of learners did a similar investigation to you and Friedrich Mohs. They tested how hard their rocks were:

Scratched by				
	Rock 1	**Rock 2**	**Rock 3**	**Rock 4**
Rock 1		✘	✘	✘
Rock 2	✔		✔	✘
Rock 3	✔	✘		✘
Rock 4	✔	✔	✔	

Record your results from Activity 1 on page 124 in a table like this.

2

The group of learners made a scale of hardness from 1–4 for their rocks. 1 was hardest, it scratched all other rocks; 4 was softest, all other rocks scratched it.

a Look at the above table. Which was the hardest rock? Which was the softest rock?

b Make a hardness scale for your results. Were your predictions correct from Activity 1 on page 124?

A diamond is the hardest rock.

Rock names

Think like a scientist!

There are lots of different types of rocks. You may know the names of some. Geologists use what they observe about the rock's properties to help them **identify** the rock.

Science words

identify
chalk
slate
volcano

1

a Compare these pictures of rocks. Describe their appearance.

b Match the rocks to the description that is similar to your own ideas.

A

① This is sandstone. It is rough and crumbly. It feels like sand. It is an orange colour.

B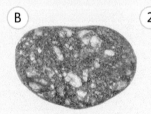

② This is granite. It has patterns and shiny crystals in it. It is hard.

C

③ This is **chalk**. It is often white. It breaks easily. It is made of very tiny pieces.

D

④ This is **slate**. It is a grey colour. It is smooth. It forms layers.

2

Obsidian, basalt and pumice are three different rocks. They come from **volcanoes**. A volcano is a mountain with a hole in the top. Very hot rocks come out of the hole.

a Look carefully at these rocks. How are they similar? How are they different?

b Find out what pumice is used for.

obsidian   basalt

pumice

How do we get rocks?

Think like a scientist!

Most rocks are under the ground. Some are under the sea. Scientists called **engineers** dig to get the rocks we need. This is called **mining**. Sometimes engineers mine for the rock in an open **quarry**. This is a big hole in the ground.

Marble quarry:
Marble is very pretty.
It is used for statues.

We can also mine in tunnels. These are dark and hot to work in.

Diamond mine: Diamonds are very hard.
They are used for jewellery.

Some rocks do not need mining. They can be found in river beds. They can be found lying on the ground.

Gold, gems and crystals can be found in rivers.

1

Do research. Find out if these rocks are quarried, tunnel mined or come from a river bed:

coal   opals   gold

limestone   marble

Let's talk

What do you think it would it be like to work in a **mine**?

Science words
engineer   mining   quarry   mine

Damage to the environment

Science words
endangered   extinct

Think like a scientist!

Plants may be killed when people start to quarry or tunnel mine. They move lots of earth and rock. More people may move to that area to do the work. This changes the environment.

It can put animals and plants in danger. We say they are **endangered**.

They can become **extinct** if their environment changes too much. Extinct means that there are no more of that animal or plant left in the world.

1

A quarry can be useful even when people have finished digging in it. Find out what we can do with quarries that we have finished digging in. Share what you find out with your class.

Let's talk

Look at the two pictures above.

a Describe how making the quarry has changed the environment.

b Would you want to live near the quarry? Why?

c Do you think it would be noisy or quiet near the quarry?

Science in context

Protect our environment

Humans affect the animals and plants around them. Sometimes they help them survive. Sometimes they do not.

Scientists study how living things live. They help us learn new things about our environment. This helps us to **protect** it.

Sometimes it is too late. Humans have made mistakes. We have harmed the living things in the environment.

The giant panda lives on bamboo. Humans are chopping down the bamboo to grow food. Why is this bad for the panda?

The pelican is covered in oil that was spilt by an oil tanker. The pelican cannot swim or fly to catch food.

Humans kill the black rhino to use its horn for medicine. There are not many rhino left. Should people be allowed to kill all the rhino?

1 Look at the pictures. Read the captions.
  a What has happened to each animal?
  b What will happen if we do not help these animals?

> **Science word**
> protect

Changes in the environment

Think like a scientist!

Scientists use HIPPO to help us understand how we change or damage an environment. HIPPO is taken from the first letter of each way humans change the environment.

(H) is for **Habitats being destroyed**. Humans chop down forests to make room for farming. We dig quarries to find rock. The animals and plants living there lose their habitats.

(I) is for **Invasion**! Humans move plants or animals to different environments. They do not belong there. They may take over the environment from the other plants and animals that normally live there.

(P) is for **People**. More and more humans are being born. People take up the space needed by animals and plants. They eat some of the food needed by other animals.

(P) is for **Pollution**. Oil spills damage the environment. Oil comes from under the sea. Humans dig this out. We use it in motor cars and for heating.

(O) is for **Overhunting**. Some people hunt animals for their skins, tusks or horns. Some people hunt plants for medicines. They hunt too much.

Science words

invasion

pollution

overhunting

1

The World Wide Fund for Nature (WWF) protects animals.

Find out what they do. Present what you find to your class.

Let's talk

Look at the animals on page 129. Which ones have been affected by which part or parts of HIPPO?

What have you learnt about planet Earth?

1

Look at these rocks. What differences can you see between them?

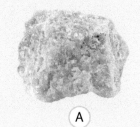 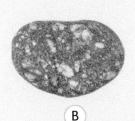  

A   B   C   D

2

Viti is not sure about the different ways that rocks can be found. Write a note telling her how we get rocks from the ground.

What can you do?

You have learnt about planet Earth. You can:
- ✔ name different rocks.
- ✔ describe the properties of rocks.
- ✔ describe what a geologist does.
- ✔ describe some ways we can get rocks.
- ✔ compare and describe rocks.
- ✔ describe how humans change the environment.
- ✔ explain some of the effects of changes to the environment.

A day on Earth

What do you remember about Earth and space?

a Do you know where you live? Do you know your address? Try writing it out. Include the country you live in. And the planet!

b Do you remember the name of our most important light source? Where can we find it?

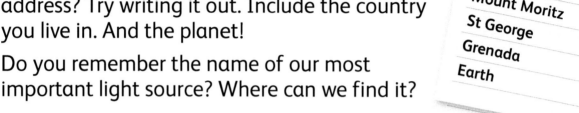

75 Willis Road
Mount Moritz
St George
Grenada
Earth

Think like a scientist!

We live on a planet that is travelling through space. It is surrounded by lots of **stars**. One star is very close to us. It gives us heat and light. Without this star, there would not be any life on planet Earth! What is this star?

Science word
stars

Let's talk

a Do you know how long a day lasts? What happens during your day?

b Talk to a partner about how you spent the day yesterday.

c Write a diary entry to record what you did.

Track the Sun

Think like a scientist!

Have you ever looked at the sky at different times of the day? The Sun appears to move across the sky. The Sun does not move, but the Earth does. But you may have noticed that the Sun appears in different places in the sky. We can track the Sun.

1

You will need…
- yellow and orange tissue paper
- sticky tape
- clock

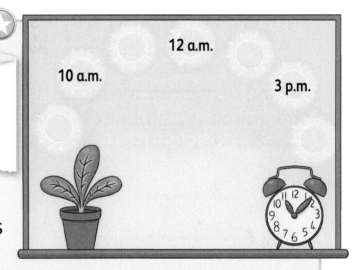

Track where the Sun appears at different times of the day. Do this investigation over two days.

a Find a sunny classroom with a window. Predict where you will see the Sun in the morning, at lunchtime and in the afternoon.

b Stick a circle of yellow tissue paper on the window to show your predictions. Write the time on each piece of paper.

c Check where the Sun is every two hours. Stick an orange circle of paper where the Sun actually is at that time.

d Were you right? Take a photograph or draw your results. Compare your results with your predictions.

Challenge yourself!

Try this at home. Does the Sun appear to move in the same way?

Work safely! ⚠

Do not look at the Sun. It will damage your eyes.

Record shadow changes

Science word
shadow

Think like a scientist!

You know that the Sun is our brightest source of light. We can use the Sun's brightness to observe where it is during the day. We can track it with **shadows**. When we stand with the Sun behind us, we can see our shadow on the ground.

Let's talk

Compare your shadows on the ground and in your photographs from Activity 1.

a What do you notice about the length and direction of your shadows as the day goes on?

b When is your shadow the shortest? Where is the Sun at this time?

c When is your shadow the longest? Where is the Sun at this time?

1

During the day, sunlight comes from different directions.

a Go into the playground in the early morning. Stand in a sunny place.

b Mark the ground with chalk to remember where you are standing.

c Look at where the Sun is in the sky.

d Ask your partner to draw around your shadow.

e Come back to the same place every hour.

f Look at where the Sun is at this time.

g Ask your partner to draw around your shadow.

h Take a photograph of your drawings at the end of the day.

Here is David's shadow drawing.

Does yours look like this?

Investigate with shadow sticks

Think like a scientist!

We can model how shadows change during the day. Shadows are made when light is blocked. When we stand in the Sun's light, we stop the light from reaching the ground. This makes a shadow. This is a place where light cannot reach.

1

You will need...
- pencil
- sticky tape or putty
- flashlight

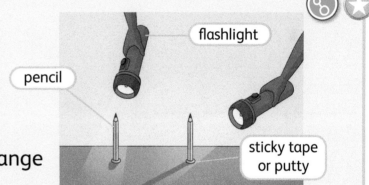

Investigate how shadows change during the day.

a Stick down your pencil with sticky tape or putty.

b Use your flashlight to make a shadow of the pencil.

c Make the shortest and longest shadows you can. Where is the light source?

d Make your pencil shadow on the right-hand side. Where is the light source?

e Predict where you must put the light source to make a shadow on the left-hand side. Test your idea.

f Compare two shadows. Draw pictures.

g Describe any pattern you observe between where you put the flashlight and what the shadow looks like. For example, as you move the flashlight to the left, what happens to the shadow?

Let's talk

Imagine that the Sun is directly above you. Predict what the shadows will look like.

What have you learnt about Earth in space?

1

Copy and complete the sentence.

Shadows are made when

_____.

2

Draw a picture of a window that the Sun shines through.

a Predict where the Sun will be at lunchtime.

b Where will it be when you leave school at the end of the day?

What can you do?

You have learnt about the Earth in space. You can:
- ✔ use a model to show how the Sun appears to move.
- ✔ describe how the model appears to move in the sky.
- ✔ use shadows to show the Sun's movement.
- ✔ explain whether it is the Sun moving or the Earth.
- ✔ say how long a day is.

1 Some learners sorted these rocks into two groups. What name would you give each group?

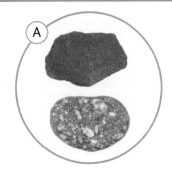

2 Complete the sentences. Use these words:

(Mohs) (scale) (geologist)

Friedrich _____ *was a scientist called*

a _____. *He compared rocks and sorted them by*

how hard they were. It is called the Mohs _____.

3 How long does it take the Earth to turn once? This is what we call one day.

4 Zara stood in the sunshine. Maris noticed her shadow. They started to record where Zara's shadow was. Complete the picture to show where Zara's shadow would be.

5 Look at this clock face.
   a Which way is clockwise?
   b What is the other direction called?

6 Look at this rock. Which of these words can we use to describe it?

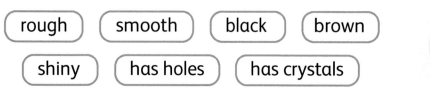

(rough) (smooth) (black) (brown)

(shiny) (has holes) (has crystals)

7 Annay put stickers on his bedroom window to show how the Sun appeared to move during the day. At what time did he put stickers A and B on his window?

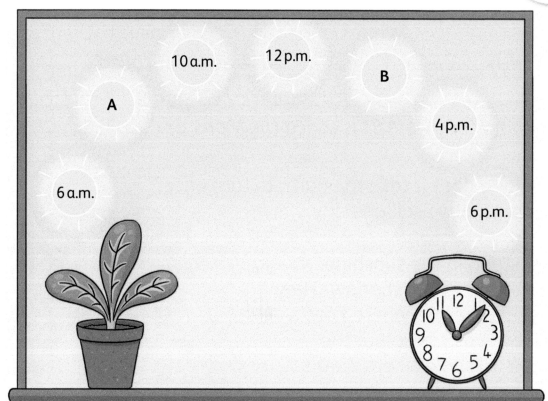

10 a.m.   12 p.m.

B

A

4 p.m.

6 a.m.

6 p.m.

8 HIPPO tells us the different ways humans can affect the environment. One P is for People taking over the environment. What does the other P stand for?

Science dictionary

A

Absence When something is missing

Absorbent A material that soaks up water easily

Absorbency To absorb something

Active To move about

Adults Grown-up humans, animals or plants

Air A gas all around us, that we cannot see, but is needed by all living things

Alive To be living

Appliance An object that uses electricity to do something, such as light up, move or make sounds

B

Balanced The different and correct types and amounts of food

Battery Two or more cells joined together

Beans Vegetables from the seed or pod of a plant

Bigger A word used to compare the size of something

Block graph A way of showing results

Blowing Air being forced or pushed out of something, such as the mouth

Body The whole of a human or other animal

Brighter A word used to compare how much light there is

Brush To move something up and down or in circles to remove dirt

C

Canine A tooth you use to grip food

Cavity A hole

Cell An object that produces electrical energy

Ceramic An object made from clay that has been shaped and baked

Chalk A type of soft, white rock

Change To make different from before

Chew To use your teeth to cut food small enough to swallow

Chickenpox A disease that is easy to catch, causes a fever and red spots on the skin

Circuit A cell, wires and other electrical components joined together so that electricity can flow in a complete circle

Classify To arrange objects in groups with similar objects

Cocoon A covering made by a silkworm

Coir A type of plant material

Compare To look at the similarities and differences between two or more objects

Complete To have no gaps in it

Component Any part in an electrical circuit

Compress To squash or squeeze something

Cooling To make something colder

Cotton A natural material made into clothes; comes from cotton plants

Crystal Shiny material found in rocks

D

Dairy Foods made from milk, which comes from cows, sheep and goats

Dark Where there is no light

Decide To choose something

Definition To explain or describe a word so that everyone knows what it means

Dentist Someone who cares for our teeth

Describe To give more details

Diagram A simple drawing to show an object or idea, with labels

Diet Everything we eat and drink

Different Not the same

Dim Where there is not much light

Direction The way something moves or the path it takes

Discovered To find something new

E

Effect The result of something; how an object moves or changes because of a force acting on it

Eggs Part of a food group with meat and fish; keeps our skin and eyes healthy

Electricians People who work with electricity

Endangered To be in danger of being harmed or lost forever

Engineer A type of scientist who designs and builds machines, engines, electrical equipment, or roads, railways and bridges

Environment The local area, weather and landscape where something lives

Equipment Tools and other objects needed for a particular task

Exercise To move about and be active

Experiment To carry out tests on something

Explorer A type of scientist who goes to new places to find things

Extinct Something that no longer exists

Extracted Taken out or removed

F

Fabric A material used to make things, such as clothes, curtains and bedsheets

False Not true or not real, artificial

Fat A food group which includes butter and cream, some meats, nuts, seeds and some fruits, such as olives and avocados

Features Things that are special about something

Fewest A word used to compare how few you have of something

Filament A thin wire in a lamp that glows so that we have light when electricity flows through it

Filling A material used to fill or to repair a hole in your teeth

Fish Part of a food group with meat and eggs; keeps our skin and eyes healthy

Flexible Bends easily; not rigid

Food Needed by all living things; gives us energy to move

Food group A group of similar foods, such as fruit and vegetables or dairy

Force A push or a pull

Freezer Where we put things to make them cold

Frozen To become hard

Fruit A food that has seeds in it, such as a tomato; it comes from a plant

Function What something does or is used for

G

Gadget A small appliance used for a particular task

Geologist A scientist who studies rocks

Germs Small living animals that cause illness and disease

Graphs A way of showing results

Grind To press something between hard surfaces to make it into small pieces or a powder

Group To sort similar things together

Grow To get bigger (a life process)

Gum disease When gums are sore and can bleed

H

Habitat The special place where an animal or plant lives

Healthy To eat and drink the right things, to exercise and move

Healthier A word used to compare how healthy someone or something is

Heat To get hot

Humans People

Hungry To need or want to eat food

Hygienic To be clean

I

Identify To recognise and name something

Illness Something that causes you to feel sick or unwell

Image A drawing, diagram or picture of something

Inactive Not active

Incisor A tooth at the front of your mouth that you use for biting

Ingredients The different parts of a mixture

Invasion To come into a place with force and in large numbers

Invent To create or design something new

Inventor A person who creates or designs a new thing

Invertebrates Animals with no backbones, such as worms

Investigate To explore, test and find out an answer to a question

L

Laboratory A room or building with scientific equipment where scientists can do tests and experiments

Landscape A large area of land with hills, valleys, rivers, trees, grass, bushes

Larvae A form of an animal that has left its egg but has not yet grown into an adult, for example, a silkworm (singular: larva)

Life processes How we can tell if something is a living thing; it needs food and air, it grows, moves and has offspring

Light What we need so that we can see

Lifestyle How active you are, what you eat and how much you sleep

Living thing A human, animal or plant that can grow, eat, move and sense things

Longer A word used to compare how long something is

M

Manufactured Made by people, not found in nature

Material What things are made of, for example, wood, fabric, plastic, leather, glass

Measure To see the exact size or amount of something

Measurement To find out how much of something there is, such as its length, weight or height

Meat Part of a food group with fish and eggs; keeps our skin and eyes healthy

Micro-habitat A very small habitat

Milk teeth The first set of teeth we have as young humans

Mind map A diagram to show the links between things

Mine A tunnel underground where rock is dug out and removed

Mining Digging out and removing something from the ground

Model Not real; helps us to understand something, or shows us a different way

Moist Damp or wet

Molar A tooth you use to grind food

Move To go from one place to another

Muscle A part of the body that helps you to move

N

Natural Something found in nature

Negative (–) One terminal on a cell; not positive

O

Object A thing we can see or touch

Observe To use our senses to find out about something

Observations The things we have observed

Offspring The young or babies of a living thing

Older A word used to compare ages of people or things

Opaque A material that is not see-through; not transparent

Opposite Heating is the opposite of cooling. Left and right are opposites. Push and pull are opposites.

Overhunting To hunt so much that you harm or endanger a plant or animal

Overlap To have parts or features that are the same

P

Parents Living things that are grown up and have offspring

Pattern A repeated design or set of numbers; we can use patterns to predict what comes next

Plaque A yellow or white sticky substance that can cause tooth decay

Plug Found on appliances that use mains electricity

Pollution Harmful things that are introduced into the environment

Positive (+) One terminal on a cell; not negative

Prediction A scientific guess of what you think will happen, using what you already know

Premolar A tooth you use for grinding food

Property The way a material behaves, such as stretchy or stiff

Protect To look after something

Pull A force that moves something towards you

Push A force that moves something away from you

Q

Quarry A large hole in the ground where rock is dug out

R

Record To write down

Research To find information by using computers and books, or by asking people

Results Information we find out after we do a test

Rigid An object that cannot be bent; not flexible

Robot A machine that can do jobs to help us

Rough A material that is bumpy, not smooth; it may also be scratchy or hard

Rubber A natural material that comes from from a tree

S

Same To match and not be different

Scale of hardness A scale to measure how hard different rocks are; invented by Friedrich Mohs

Scales The skin covering of a fish

Screen time The time we use to watch television or play computer games

Senses We have five of these: sight (seeing); hearing; smelling; tasting and touching (feeling)

Shadow A dark shape made when an object stops or blocks light

Shallow A short distance to the bottom; not deep

Shelter A place in which something lives, that gives safety and protection

Silk A type of cloth; a natural material made by silkworms

Similar People or things that are almost the same, but are not exactly the same

Similarities The things that are almost the same

Skin covering What all animals are covered in, for example scales, fur or feathers

Slate A dark, grey rock

Sleep The time when our body rests and can repair itself

Smooth A material that is even with no bumps; not rough

Socket Where an appliance goes into the plug to get mains electricity

Soft A material that is not hard or rough; can feel nice to touch

Soggy Wet and soft

Soil A mixture of tiny pieces of rock and dead plants

Sort To put into groups or classify

Source Where something comes from

Space Anything beyond the Earth; the area where the planets and stars are found

Stars The lights you can see in the sky at night; huge, spinning balls of hot gases that are very far away

Stone A small piece of rock

Strong Something that does not break easily

Sugar Found in many foods which give us energy; too much can cause tooth decay

Suitable Right for its job or purpose

Sun The most important source of light

Surface The top part of something

Survey A way of finding out answers to questions from lots of different people

Swallow When the food you are chewing goes from your mouth and into your stomach

Symbols Images or pictures that give you information quickly

Symptoms Signs that tell us information about what illness we have

T

Taller A word used to compare the height of something

Tally To record or count something

Temperature To measure how hot someone or something is

Terminal The end of something, such as a cell

Test An experiment to find an answer

Texture A property of a material; how something feels

Thirsty To need water to drink

Tooth decay When teeth start to go bad

Toothbrush A brush you use to clean your teeth

Toothpaste A substance that you use to clean your teeth

Transparent A material that is clear and see-through; not opaque

U

Useful Something that helps you to do something more easily

V

Vegetable A food that comes from a plant, such as a potato

Venn diagram A way of showing things that have been grouped and sorted; using sorting circles

Veterinarian (vet) A person who cares for animals

Volcano A mountain with a hole in the top where very hot rock comes out

W

Water A clear liquid (that we all need to stay alive)

Waterproof A material that does not let water through

Y

Young The children or babies of a living thing